DEFECT CRYSTAL CHEMISTRY
AND ITS APPLICATIONS

R.J.D. Tilley
Professor of Materials Science
University College, Cardiff

Blackie
Glasgow and London
Published in the USA by
Chapman and Hall
New York

Blackie & Son Limited,
Bishopbriggs, Glasgow G64 2NZ
7 Leicester Place
London WC2H 7BP

Published in the USA by
Chapman and Hall
in association with Methuen, Inc.
29 West 35th Street, New York, NY 10001

British Library Cataloguing in Publication Data

Tilley, R.J.D.
Defect crystal chemistry and its
applications.
1. Point defects   2. Chemistry, Inorganic
I. Title
546   QD921
ISBN 0-216-92003-5

Library of Congress Cataloging-in-Publication Data

Tilley, R.J.D.
Defect crystal chemistry and its applications.
Includes bibliographies and index.
1. Crystals—Defects   2. Point defects   I. Title
QD921.T53   1986   548'.842   86-989
ISBN 0-412-01331-2

Photosetting by Thomson Press (I) Ltd., New Delhi
Printed in Great Britain by Bell & Bain (Glasgow) Ltd.

Preface

To cover all aspects of the chemistry and physics of defects in all types of solids is an immense task. Nevertheless the subject is of increasing importance in many areas of science and engineering, and students need to be aware of the role of defects in controlling the chemical and physical properties of the solid materials in common use today.

This book arises from courses given to advanced undergraduates and immediate postgraduates in chemistry, physics and materials science. It introduces newcomers to the subject with particular reference to non-stoichiometric crystals. Part 1 considers the rather idealized situation of point defects in crystals which show no composition range at all, emphasizing the influence of the defects on crystal properties and the way in which the property variations can be utilized in practical devices.

As the reader becomes more familiar with the introductory concepts, the topics are broadened to cover our present understanding of the nature of non-stoichiometric solids. Part 2 deals with materials containing ions with fixed valence, and Part 3 extends this treatment to materials containing atoms of variable valence. Part 4 provides a review of the complex and beautiful structures to be found in real non-stoichiometric compounds, followed by a consideration of the way in which current theory attempts to account for these structures in terms of the concepts with which the book opens. The aim throughout is to provide an introduction which will serve as a framework for further discussion. A list of source books and additional reading appears at the end of each chapter, including references to the original literature.

It gives me great pleasure to acknowledge the debt that I owe to many colleagues who have helped me over the years to understand the subject matter that has formed the basis of this book. In particular my thanks are due to J.S. Anderson, Sten Andersson, D.J.M. (Judge) Bevan, LeRoy Eyring, B.G. Hyde, E. Iguchi and F.S. Stone. Naturally, I take full responsibility for any errors in the scientific content. It is a pleasure to thank Mrs J. Jones for drawing the figures. Finally I wish to acknowledge my indebtedness to my wife and family for their encouragement and tolerance, without which this book would never have been written.

RJDT

For Anne, Elizabeth, Gareth and Richard

Contents

Part 2   Non-stoichiometric materials containing ions with fixed valence

4 Non-stoichiometry and defect chemistry   75

5 Some applications: galvanic cells and sensors using non-stoichiometric compounds   99

Part 3   Non-stoichiometric materials containing atoms of variable valence

6 Non-stoichiometry and electronic conduction in materials containing ions of variable valence   119

Part 1 Stoichiometric crystals

1 Point defects in stoichiometric crystals

1.1 Introduction

During the course of this century, and particularly in more recent years, it has been realized that many properties of solids are controlled not so much by the structure of the material itself but by faults or defects in this structure. Thus the strength of metals is often governed by the presence of linear defects called dislocations, and the task of hardening or strengthening metals has much to do with either locking the dislocations into the structure so that they cannot move, or else eliminating them as far as possible. Similarly, the various and beautiful colours of many gemstones are due to impurity atoms within the otherwise perfect structure of the crystal itself. Boundaries between the crystallites which compose a polycrystalline solid also play a significant role in controlling the strength of such materials and in determining the way the solid behaves when an electric field is applied across the material. Indeed there is no aspect of the physics and chemistry of solids which is not decisively influenced by the defects that occur in the material.

This book relates to a small part of this large subject area. It is concerned mainly with some of the chemical and physical effects of the presence of mistakes in the crystal. We commence with the simplest case, mistakes at isolated or fairly isolated atomic sites. They are, because of this, termed *point defects*, as they generally involve the replacement or removal of single atoms at discrete points within the crystal structure.

These defects affect the chemical properties of solids in very significant ways, and play a key role in controlling the rate at which solids react. A knowledge of point defect behaviour is therefore fundamental to an understanding of a wide variety of solid state reactions of importance, including sintering, fabrication of new phases via solid state reaction, and corrosion. They also play an important role in controlling the physical properties of many materials, particularly the electronic properties of semiconductors and the colours of insulators. In this and subsequent chapters we will discuss how these defects are able to exert such an important role on the chemical and physical properties of solids.

In order to do this we need to invent some structural pictures of point defects. At the simplest level we can imagine that, in a normal crystal, an atom will sometimes be absent from a normally occupied position, to create a *vacancy*; sometimes an atom will take up a position in the crystal which is not a normally occupied site, to create an *interstitial atom*, and sometimes atoms will be present on sites normally associated with a different atom type. These atoms may be impurities, or else may simply be one of the component atoms in the structure which has somehow become misplaced from its proper location.

Of course, these elementary defect types interact in a wide variety of ways, both with each other, with dislocations, stacking faults and grain boundaries in crystals, and also with any electrons that may be present. We can thus build up a complex defect chemistry in which defect aggregation and clustering become important and dominate the physical or chemical behaviour of the material. Such interactions are the norm in inorganic crystals and will be considered in some detail later in this volume. At the outset, though, we will consider the sort of point defect populations that we would reasonably expect to find in a pure crystal, an oxide or a fluoride, say, where the composition of the crystal is fixed and the concentration of point defects is low. This will then act as a platform to underpin discussion and understanding of the more complex situations that exist in most systems accessible to experimental investigations. Our first task, therefore, is to envisage a simple model of a point defect in a crystal and try to estimate what sort of defect populations we would expect in these idealized circumstances.

1.2 The equilibrium concentration of point defects in crystals

1.2.1 *Schottky defects*

One of the simplest point defects that we can imagine to occur in a crystal is a vacancy. In general, though, we cannot simply introduce vacancies in any number or fashion into the crystal or else the charge and mass balance will be upset. We therefore need to be sure that the numbers are correctly balanced so as to maintain the stoichiometric formula and preserve electrical neutrality. Such a situation was envisaged by W. Schottky and C. Wagner, whose ideas were first presented in 1930. The defects arising from balanced populations of cation and anion vacancies are now known as *Schottky defects*. For example, if the crystal has a formula MX, then the number of cation vacancies will be equal to the number of anion vacancies, in order to maintain electrical neutrality. In such a crystal one Schottky defect consists of one cation vacancy and one anion vacancy, although these vacancies are not necessarily imagined to be asssociated with each other in any way. It is necessary to remember, therefore, that the number of Schottky defects in a crystal of formula MX is equal to one half of the number of vacancies. In a crystal of formula MX_2 there will be twice as many anion vacancies as cation vacancies, again to preserve charge neutrality, and similar considerations apply to compounds with other compositions. In these cases one Schottky defect will consist of the appropriate numbers of cation and anion vacancies to form an electrically neutral total.

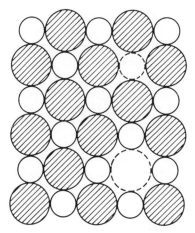

Figure 1.1 Schematic illustration of a Schottky defect in a crystal of composition MX. The metal atoms are drawn smaller than the non-metal atoms and are unshaded in this illustration. The defect consists of vacancies, shown here as dotted circles, on both metal and non-metal sites.

Schottky defects are frequently represented diagrammatically by a drawing of the sort shown in Figure 1.1. This represents a Schottky defect in a crystal structure of the sodium chloride type, the projection shown in this diagram representing a {100} plane. While such diagrams have a limited use and you may find them of some help, it is important to remember that they are gross simplifications. In order to gain an understanding of the structural nature of such defects it is preferable to try to build crystal models. In addition, you should try to imagine Schottky defects in real crystals of various structure types, and not stick to the NaCl type of structure illustrated here. When doing this, do not forget that in real crystals the atoms are vibrating, and that they will move towards or away from a vacancy if one is created, depending upon the bonding in the crystal under consideration. This process is called *relaxation.* Moreover, remember that it will be much easier to create vacancies at some atom sites than others, a fact that is not apparent from diagrams.

Having thus constructed a structural model for Schottky defects, we now want to show that they can be in thermodynamic equilibrium in a crystal and so will be present under normal circumstances. The usual way to tackle this is to start by considering the Gibbs' free energy of a crystal, which is written as

$$G = H - TS \qquad (1.1)$$

where G is the Gibbs' free energy, H is the enthalpy, S is the entropy and T the absolute temperature of the crystal. If we introduce Schottky defects, we introduce a change in the free energy of the crystal by an amount ΔG, given by

$$\Delta G = \Delta H - T\Delta S \qquad (1.2)$$

where ΔH is the associated change in enthalpy and ΔS the change in the entropy of the crystal. At equilibrium, the free energy will be a minimum. The problem is to estimate this point theoretically, and to show that it occurs when we have a finite population of defects present.

To attempt this, we can try to estimate the change in the enthalpy ΔH, the entropy ΔS, or both. The enthalpy tends to be associated more with nearest neighbours and the bonding energy between them and is not easy to assess theoretically. The change in entropy, ΔS is also complex, and consists of terms due to the vibration of the atoms around the defects and terms due to the arrangements of the defects in the crystal. This latter quantity, called the *configurational* entropy, is relatively easy to assess using the well established methods of statistical mechanics, and this is what we will estimate here.

For illustration we will consider a crystal of overall composition MX. In such a crystal, suppose n_s is the number of Schottky defects per cm^3 in the crystal at $T^\circ K$, that is, we have n_s vacant cation sites and n_s vacant anion sites present. In addition in a crystal of this type there are N possible cation sites and N possible anion sites per cm^3. We can determine the entropy change, ΔS, in a system of occupied and unoccupied sites by using the Boltzmann equation

$$S = k \ln W \qquad (1.3)$$

where S is the entropy of a system in which W is the number of ways of distributing n defects over N sites at random and k is Boltzmann's constant. Probability theory shows that W is given by the formula

$$W = N!/(N - n)!n! \qquad (1.4)$$

where the symbol $N!$, called factorial N, is mathematical shorthand for the expression

$$N \times (N - 1) \times (N - 2)...1. \qquad (1.5)$$

Returning to our case, the number of ways that we can distribute the n_s cation and anion vacancies over the available sites in the crystal will be given by

$$w_c = N!/(N - n_s)!n_s! \qquad (1.6)$$

for vacancies on cation sites, and

$$w_a = N!/(N - n_s)!n_s! \qquad (1.7)$$

for vacancies on anion sites. For a crystal of stoichiometry MX,

$$w_c = w_a. \qquad (1.8)$$

The total number of ways of distributing these defects, W, is given by the product of w_c and w_a, hence

$$w = w_a w_c = w^2 \qquad (1.9)$$

Therefore the change in entropy caused by introducing these defects is

$$\Delta S = k \ln (w^2) = 2k \ln w \qquad (1.10)$$

i.e.
$$\Delta S = 2k \ln [N!/(N - n_s)!n_s!] \tag{1.11}$$

Now this expression, as it stands, is of little help to us in evaluating ΔG, and it must be simplified somewhat to be of use. In fact we need to eliminate the factorials. This is usually done by employing the approximation

$$\ln N! \approx N \ln N - N \tag{1.12}$$

which is invariably referred to as Stirling's approximation. In fact the approximation in equation (1.10) is not particularly good and is several percent in error even for values of N as large as 10^{10}. The correct expression for Stirling's approximation is

$$\ln N \approx N \ln N - N + 1/2 \ln (2\pi N) \tag{1.13}$$

which is accurate even for very low values of N. Nevertheless, in order to continue without using excessively cumbersome mathematical expressions we revert to the simpler expression given in equation (1.12). Substituting into equation (1.11) we ultimately obtain

$$\Delta S = 2k\{N \ln N - (N - n_s) \ln (N - n_s) - n_s \ln n_s\} \tag{1.14}$$

We make no attempt to calculate the enthalpy change, but merely label the enthalpy needed to form a Schottky defect ΔH_s. To form n_s pairs we need a total enthalpy input of $n_s \Delta H_s$, hence

$$\Delta G = n_s \Delta H_s - 2kT \{N \ln N - (N - n_s) \ln (N - n_s) - n_s \ln n_s\} \tag{1.15}$$

In general, the energy increase due to the ΔH_s term will be offset by the energy decrease due to the $-\Delta S$ term. At equilibrium we can write

$$(d\Delta G/dn_s)_T = 0 \tag{1.16}$$

i.e.
$$(d\Delta G/dn_s)_T = d/dn_s\{n_s \Delta H_s - 2kT[N \ln N$$
$$- (N - n_s) \ln (N - n_s) - n_s \ln n_s]\} = 0 \tag{1.17}$$

Remembering that $N \ln N$ is constant, so its differential is zero, and that the differential of $\ln x$ is $1/x$ and of $x \ln x$ is $1 + \ln x$, we find, on differentiating,

$$\Delta H_s - 2kT \, d/dn_s[N \ln N - (N - n_s) \ln (N - n_s) - n_s \ln n_s] = 0 \tag{1.18}$$

i.e.
$$\Delta H_s - 2kT[\ln (N - n_s) + (N - n_s)/(N - n_s) - \ln n_s - n_s/n_s] = 0 \tag{1.19}$$

hence
$$\Delta H_s = 2kT \ln [(N - n_s)/n_s] \tag{1.20}$$

Rearranging,
$$n_s = (N - n_s)e^{-\Delta H_s/2kT} \tag{1.21}$$

or, if N is considered to be very much greater than n_s,

$$n_s \approx Ne^{-\Delta H_s/2kT} \tag{1.22}$$

In using this equation, remember ΔH_s is the enthalpy to form one defect. The

Table 1.1 The formation enthalpy of Schottky defects in some compounds of formula MX

Compound	$H_s (J) \times 10^{-19}$
MgO	10.574
CaO	9.773
SrO	11.346
BaO	9.613
LiF	3.749
LiCl	3.397
LiBr	2.884
LiI	2.083
NaCl	3.685
NaBr	2.692
KCl	3.621
KBr	3.797
KI	2.563
CsBr	3.204
CsI	3.044

units of ΔH_s are thus joules per defect or vacancy pair, and k, Boltzmann's constant, is in $J\,deg^{-1}$. In the literature the units used are usually expressed in molar quantities. Hence ΔH_s is given as $J\,mol^{-1}$. In this case equation (1.22) becomes

$$n_s \approx Ne^{-\Delta H_s/2RT} \tag{1.23}$$

where ΔH_s is in $J\,mol^{-1}$ and is the energy required to form one mole of Schottky defects; R, the gas constant, is equal to kN_A where N_A is Avogadro's number, $6.0225 \times 10^{23}\,mol^{-1}$, so that R has units of $JK^{-1}\,mol^{-1}$. Some experimental values for the enthalpy of formation of Schotty defects are given in Table 1.1.

To obtain an idea of the number of Schottky defects which are likely to occur in a crystal we have computed n_s/N and n_s from equation (1.23) taking ΔH_s to have a value of $200\,kJ\,mol^{-1}$, which is a reasonable value for a typical alkali halide. We find the results set out in Table 1.2. The values are of interest, as they show that the population of point defects is quite low. Even at 900 K we find that only one or two sites in one million are vacant.

It is sometimes useful to make a rough estimate of the fraction of sites in a crystal which are vacant due to Schottky disorder. This figure can be obtained by the following procedure.

Table 1.2 Schottky defect concentrations in MX compounds

Temperature (°C)	Temperature (K)	n_s/N	$n_s(cm^{-3})$
27	300	3.87×10^{-18}	7.26×10^4
127	500	3.57×10^{-11}	6.67×10^{11}
427	700	3.45×10^{-8}	6.47×10^{14}
627	900	1.57×10^{-6}	2.94×10^{16}

Taking logarithms in equation (1.23),

$$\ln n_s = \ln N - \Delta H_s/2RT \qquad (1.24)$$

and, as we know that

$$\ln x = 2.3026 \log_{10} x \qquad (1.25)$$

it is possible to write

$$\log_{10} n_s - \log_{10} N = -\Delta H_s/2.3026 \times 2RT \qquad (1.26)$$

hence

$$\log_{10} (n_s/N) = -\Delta H_s/4.6052RT \qquad (1.27)$$

and substituting for R a value of $8.3143 \, J \, K^{-1} \, mol^{-1}$ and removing the logarithm term, we obtain

$$n_s/N \approx 10^{-\Delta H_s/39T} \qquad (1.28)$$

where ΔH_s is measured in $J \, mol^{-1}$.

Remember that this formula only applies to materials with a composition MX, as equation (1.23) was the starting point of the analysis.

1.2.2 Frenkel defects

Frenkel defects, like Schottky defects, also involve vacancies in the crystal structure. In this case, though, the vacancies only exist on one sub-lattice and the atoms which should occupy these vacant positions are placed in alternative sites which are normally unoccupied. Such a site is called an *interstitial* position. Hence, if we consider a crystal of formula MX, a Frenkel defect consists of one interstitial atom plus one vacant site in the position where that atom would normally be found. We can illustrate this schematically (Figure 1.2). However, one should take care to remember that, as with Schottky defects, this illustration is highly idealized when compared to the structure of a defect in a real crystal.

The calculation of the number of Frenkel defects in a crystal proceeds along lines parallel to those for Schottky defects. Suppose there are N lattice sites per cm^3 in the array of atoms affected by Frenkel defects, and N^* available interstitial sites. If n_f ions from the lattice move into interstitial sites, each needing an enthalpy ΔH_f, the total enthalpy change is given by $n_f \Delta H_f$. This quantity is not easily calculated theoretically, and, as before, we turn to a assessment of the configurational entropy of these vacancies and interstitial atoms in order to proceed further.

In an analogous way to our discussion of Schottky defects, we can write down the number of ways of distributing the vacancies over the available positions in the atom array affected by Frenkel defects as

$$w_v = N!/(N - n_f)! \, n_f! \qquad (1.29)$$

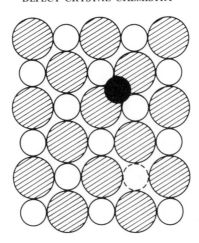

Figure 1.2 Schematic illustration of a Frenkel defect in a crystal of composition MX. The metal atoms are drawn smaller than the non-metal atoms and are unshaded in this illustration. The defects consist of equal numbers of vacancies on either the metal or non-metal sub-lattice and interstitial ions of the same type. Here the vacancy, shown as a dotted circle, is supposed to be on the metal sub-lattice. The interstitial metal atom is shown shaded.

where we have n_f vacancies and a possible total of N positions for the location of the vacancy. Similarly, for the distribution of the interstitial atoms we can write

$$w_i = N^*!/(N^* - n_f)!\,n_f! \qquad (1.30)$$

Proceeding in exactly the same way as for Schottky defects, we can write the total number of ways of arranging the vacancies and interstitials as W, where

$$W = w_v w_i \qquad (1.31)$$

The change in configurational entropy, ΔS, due to this distribution will be given by

$$\Delta S = k \ln w_v w_i \qquad (1.32)$$

Hence

$$\Delta S = k\{\ln [N!/(N - n_f)!\,n_f!] + \ln [N^*!/(N^* - n_f)!\,n_f!]\} \qquad (1.33)$$

Once again we have recourse to Stirling's theorem to put this into a more useful format for our needs. This procedure ultimately yields the cumbersome expression

$$\Delta S = k[N \ln N + N^* \ln N^* - (N - n_f) \ln (N - n_f) \\ - (N^* - n_f) \ln (N^* - n_f) - 2n_f \ln n_f] \qquad (1.34)$$

Note that if we make N^* and N equal, we arrive once again at the expression for Schottky defects. Proceeding as before, the free energy change ΔG_f to form the defects is given by

$$\Delta G_f = n_f \Delta H_f - kT [N \ln N - N^* \ln N^* - (N - n_f) \ln (N - n_f)$$
$$- (N^* - n_f) \ln (N^* - n_f) - 2n_f \ln n_f] \tag{1.35}$$

Setting $(d\Delta G_f/dn_f)_T = 0$ and differentiating, remembering that $N \ln N$ and $N^* \ln N^*$ are constants and so are eliminated, we arrive at an expression similar to the Schottky expression:

$$\Delta H_f = kT \ln [(N - n_f)(N^* - n_f)]/n_f^2 \tag{1.36}$$

i.e.

$$n_f^2 = (N - n_f)(N^* - n_f)e^{-\Delta H_f/kT} \tag{1.37}$$

This expression can be further simplified if we make yet another approximation, and suppose the number of Frenkel defects, n_f, is much smaller than either the number of normal positions (N) or the number of interstitial positions available (N^*). In this case we can write the approximate expression

$$n_f \approx (NN^*)^{1/2}e^{-\Delta H_f/2kT} \tag{1.38}$$

As one would expect, this is similar to the Schottky formula derived in the previous section. Hence, values for ΔH_f are usually quoted in J mol^{-1}, and one therefore uses the expression

$$n_f \approx (N N^*)^{1/2}e^{-\Delta H_f/2RT} \tag{1.39}$$

where R, the gas constant, is in J mol^{-1}. Some experimental values are given in Table 1.3.

1.2.3 Further considerations

The formulae derived above apply to stoichiometric materials of formula MX. In order to discuss crystals of different stoichiometry, such as M_2X_3, MX_2 and so on, it is important to bear in mind that different, though similar, formulae will result. This arises because the configurational entropy term will involve different numbers of defects than in the more simple case of MX materials.

Table 1.3 The formation enthalpy of Frenkel defects in some compounds of formula MX and MX_2

Material	$H_f(J) \times 10^{-19}$
UO_2	5.448
ZrO_2	6.569
CaF_2	4.486
SrF_2	1.122
$AgCl$	2.564
$AgBr$	1.923
$\beta\text{-}AgI$	1.122

The formulae arrived at indicate that at all temperatures above 0 K we should expect to find point defects present in pure crystals. For this reason such defects are also termed *intrinsic* defects. Moreover, these defects will be in thermodynamic equilibrium, and so will not be removed by annealing or other thermal treatments. The actual type of defect found, either Schottky or Frenkel, will depend, in the main, on the value of ΔH. One would not expect this to be the same for these two alternatives, and hence one would anticipate that only the defect with the lower value of ΔH would predominate.

Despite the utility of these formulae, there are a number of approximations which have been made in their derivation apart from the mathematical short-cuts noted, and it is useful to mention these so that the limitations implicit in the present treatment are also apparent. The most important are:

(i) Only one sort of defect is supposed to be found in a crystal. This clearly need not be so. Indeed, the sort of defect predominating can change with the defect concentration present and at times we can assume that we have both types of defect present.

(ii) The treatment assumes that the defects do not interact. As we shall see later, this is unlikely to be true except when the number of defects present is very small. Defect interactions are important and it is possible to take such interactions into account in more general formulae.

(iii) The important quantities ΔH and ΔS are assumed to be temperature-independent. This is not likely to hold in practice; in particular, the vibrational component of the entropy, which has been neglected altogether, will become increasingly important at high temperatures. This is just the domain where the defect populations also start to become significant, of course.

These three points show the direction in which the simple theories outlined above can be modified to present a more realistic model of a solid. The interactions between defects can be calculated using a variety of more complex theories, such as the Debye–Hückel treatment used for electrolytes, and more complex computational techniques which rely on the power of modern computers for speed. We will return to this latter topic later in this book. Different ways of distributing defects over the available lattice positions in a crystal can be envisaged, and ways to estimate the entropy of such distributions can also be sought. This approach can also include more sophisticated site exclusion rules, which allow defects to either cluster or keep apart from each other.

Perhaps more important than these points is the fact that we have totally ignored the role of impurities or deliberate additives in influencing the number or types of defect present in a crystal. These matters, however, are best treated in respect of non-stoichiometric solids and will be considered further in later chapters.

1.3 The photographic process

1.3.1 *Light-sensitive crystals*

Photography is one of the most widely used of all information storage methods. The modern processes of both black-and-white and colour photography rely upon the same fundamental point defect chemistry for their operation and it is this that we will discuss briefly in this section.

The light-sensitive materials employed in photography are silver halides, notably AgBr or a mixture of AgBr and AgI. These are dispersed in gelatin to form the emulsion. In practice, the silver halide crystals are grown within the gelatin matrix so as to make them as free of macroscopic defects such as dislocations as possible, as these have a deleterious effect on the perfection of the photographic images ultimately produced. The crystals so formed are usually thin triangular or hexagonal plates, varying between 0.01 and 10 μm in size, and in photographic parlance are known as 'grains'. Not all defects are undesirable, though, and point defects, which we have seen will always be present in the crystals, are themselves responsible for the photographic process.

When the emulsion is exposed to light a *latent image* is said to form. After illumination, each grain will either contain a latent image, that is it will have interacted with the light photons, or it will have remained unchanged. The film is then put into a developer. Each grain which contains a latent image is totally reduced to metallic silver. Each crystallite with no latent image remains unchanged. The reactions taking place can be written down schematically as

$$AgBr \xrightarrow{h\nu} [AgBr + \text{latent image}] \tag{1.40}$$

$$[AgBr + \text{latent image}] \xrightarrow{\text{develop}} Ag \tag{1.41}$$

It has been found that only a few photons are needed to form the latent image, which appears to consist of a cluster of about three of four atoms of silver. As a fully developed crystallite may consist of 10^9 silver atoms, we see that the film is a very sensitive light detector.

1.3.2 *Mechanisms of latent image formation*

The silver halide AgBr, which is the material of most interest in photography, is of the NaCl structure type. Whilst in most crystals with the NaCl structure Schottky defects are the major structural point defect type, it is found that the silver halides, including AgBr, favour Frenkel defects. The enthalpy of formation of a Frenkel defect in AgBr is about 2×10^{-19} J and so we can estimate the number of Frenkel defects in a crystal of AgBr using the formula given in equation (1.38).

Despite the fact that the photographic process has been very widely investigated both experimentally and theoretically, there are still a number of questions related to the formation of latent images that need to be answered. However, some aspects are clear. It appears that the formation of latent images is a multi-stage process, involving the Frenkel defect population. The major steps are believed to be as follows.

(i) A light quantum interacts with the crystal to generate an electron. This almost certainly comes from a halogen ion. The electron is free to move in the lattice and migrates to an interstitial silver ion which is part of a Frenkel defect, to form a silver atom:

$$Ag_i^{\cdot} + e' \rightarrow Ag_i^{x} \qquad (1.42^*)$$

(ii) In many instances the above reaction will then take place in the reverse direction, and the silver atom will revert to the normal stable state as a Frenkel defect. However, the metal atom seems to be stabilized if another photon activates a nearby region of the crystal before the decomposition can take place. This stabilization may take place in either of two ways. The first of these is that the silver atom formed by the reaction given in equation (1.42) can trap the electron liberated by the second photon, thus:

$$Ag^{x} + e' \rightarrow Ag_i' \qquad (1.43)$$

The silver ion produced in this reaction is then neutralized by association with another interstitial silver atom, as shown in equation (1.44):

$$Ag_i' + Ag_i^{\cdot} \rightarrow 2Ag_i^{x} \qquad (1.44)$$

to produce a cluster of two neutral silver atoms.

The second possibility is that the second electron could interact with an interstitial ion to yield a second silver atom which would then diffuse to the first silver atom to form an identical cluster of two, viz.:

$$Ag_i^{\cdot} + e' \rightarrow Ag^{x} \qquad (1.45)$$

$$Ag_i^{x} + Ag_i^{x} \rightarrow 2Ag_i^{x} \qquad (1.46)$$

It seems that a minimum of four silver atoms is needed before the latent image is formed, and these could form by a continuation of the reactions just described.

1.4 Photochromic glass

Photochromic glass is another material which is sensitive to light. Although many types of photochromic glass have been fabricated, the best known are those which darken on exposure to visible or ultraviolet light and regain their transparency when the light is removed. Such glasses are widely used in sunglasses, automobile sunroofs and for many architectural purposes.

*The terminology used in this equation and later in this chapter to describe the defects will be explained in Chapter 4.

The mechanism of the darkening transformation is very similar to that involved in the photographic process and, in the same way, many aspects of the process are still incompletely understood. Here, therefore, we will outline the principles of photochromic behaviour and refer the interested reader to the references on this topic given in section 1.7 (Supplementary reading).

Photochromic glasses are multiphase materials which usually contain silver halides as the light-sensitive medium. The glass for this use would typically be an aluminoborosilicate (Pyrex type) material containing about 0.2 wt% of silver bromide or chloride. When the glass is first fabricated, the cooling rate is deliberately adjusted to be high. Under these conditions the silver halides remain dissolved in the glass matrix and the glass produced is transparent and does not show any photochromic behaviour at all.

This glass is transformed into the photochromic state by annealing under carefully controlled conditions of temperature and time, which might be, for example, 550 °C for 30 minutes followed by 650 °C for 30 minutes. The heat treatment is chosen so that the silver halides crystallize in the glass matrix. Care must be taken to ensure that the crystals do not become too large and that they do not aggregate. A desirable size would be about 10 nm diameter and the individual crystallites should be about 100 nm apart. These crystallites are precipitated in the complete absence of light. Following this treatment a finished glass blank will look clear, because the silver halide grains are so small that they do not scatter light appreciably.

The influence of light causes changes similar to those occurring in a photographic emulsion. The photons liberate electrons from the halide ions and these are trapped by interstitial silver ions, which exist as Frenkel defects, to form metallic silver. The detailed equations for this process have been given in the previous section. Here we can summarize them as

$$2AgX \rightleftharpoons 2Ag + X_2 \qquad (1.47)$$

where X stands for the halide atoms. The clusters of silver appear black and cause the darkening of the glass which is apparent to the eye.

In a photographic emulsion the halide molecules produced in this way can diffuse away from the silver and the process becomes irreversible. In a glass, however, the halide remains trapped near to the silver particles. This means that the silver particles can re-react with the halide molecules to form the AgX salt once again. The two reactions are taking place simultaneously under normal circumstances, so that when the amount of incident light is high we have a large number of silver particles present in the glass and hence a high degree of darkening. When the light intensity falls the number of silver particles decreases and the glass becomes less dark.

For commercially useful materials the speed of this reaction is important. If the darkening takes place too slowly, or if the subsequent fading of the colour is too slow, the materials will not be as useful as one would desire. Because of this, large numbers of different glass preparations have been devised. The most generally successful of these uses a copper halide as an additive. The role of the copper in enhancing the speed of reaction is due to the fact that it is able to exist in two valence states. It is thus able to supply the electrons that are needed in the reaction to form the silver atoms rather more easily than the halogen ions. The use of such impurities to enhance the reactivity or properties of solids is a feature that we will meet often in later chapters of this book.

1.5 The lithium iodide battery

In the two preceding sections we have looked at the operation of some materials which function because of the Frenkel defects which occur in their structures. We will now briefly examine one application of the compound lithium iodide, in which Schottky defects play an important role.

LiI batteries are widely used as the power supply for heart pacemakers which are directly implanted into the body. In these devices the LiI is the electrolyte which separates the anode and cathode of the cell. Although LiI has only a low ionic conductivity, this disadvantage is more than offset by a number of advantages which make these batteries ideal for medical use.

How is such a cell constructed? The anode is made of lithium metal, which can readily conduct the electrons flowing in the external circuit, and which do the useful work of the cell. For the cathode a complex of iodine and polyvinyl pyridine, a conducting polymer, is employed, because iodine itself is not a good electronic conductor. The cell is fabricated by placing the Li anode in contact with the polyvinyl pyridine–iodine polymer. The lithium, being a reactive metal, immediately combines with the iodine in the polymer to form a thin layer of LiI, which forms the electrolyte. For heart pacemakers, the battery itself is typically constructed of two cells, placed back to back, separated by a nickel gauze, and contained in a stainless steel or titanium case, as shown schematically in Figure 1.3a. For conventional use in electronic circuits, a single button cell is more often employed as shown in Figure 1.3b.

Under normal circumstances, after initial formation of a thin layer of LiI, the reaction would cease. However, as LiI contains Schottky defects in its structure, the small Li^+ ions can be transported across the LiI layer via the vacancy population that exists in the cation sub-lattice. On providing an external circuit, the Li atoms in the anode become Li^+ ions at the anode–electrolyte interface. These diffuse through the LiI to reach the iodine in the cathode. The route that these ions take involves them in traversing the path mapped out by the point defects in the crystals. The electrons lost by the Li in becoming Li^+ ions traverse the external circuit and arrive at the interface between the cathode and the electrolyte. Here they can change I atoms into I^- ions which then combine with the Li^+ ions to form more LiI. During use, the thickness of the LiI electrolyte gradually increases because of this reaction, and this ultimately causes the cell to become unusable.

(a)

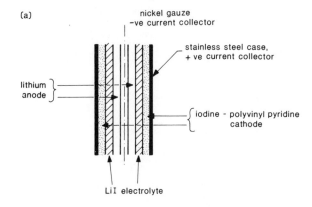

LiI electrolyte

(b)

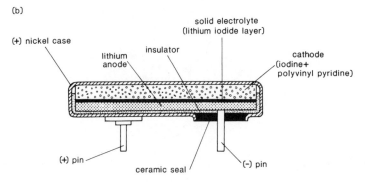

Figure 1.3 (a) Schematic construction of a back-to-back LiI cell used in heart pacemakers. (b) Diagram of a single LiI cell for use in electronic circuits.

1.6 Disordered cation compounds

In the preceding sections we have concentrated upon normal compounds in which the concentration of point defects was low, and could be reasonably approximated by the formulae given in equations (1.23) and (1.39). In this final section we draw attention to a group of unusual compounds which can be thought of as containing very high concentrations of Frenkel defects. They are not altogether new, though, and some were investigated by Faraday in the 19th century. The presently known materials which comprise the group are listed in Table 1.4.

In order to understand why these compounds can possess such high defect concentrations, let us consider an apparently simple material, AgI, which we have already mentioned in our discussion of the photographic process. In 1914 silver iodide was discovered to have a high temperature form, α-AgI, which occurs above 147 °C and which possesses an unusually high ionic conductivity for a pure, stoichiometric phase.

The structure of the phase, shown in Figure 1.4, reveals that the iodine

DEFECT CRYSTAL CHEMISTRY

Table 1.4 Disordered cation compounds related to α-AgI

α-AgI	α-CuI	Na_2S
α-Ag_2S	α-Cu_2Se	MHg_4I_5
α-Ag_2Te		(M = Rb, K, Cs)
α-Ag_2Se		
α-Ag_3SI		
α-Ag_3HgI_4		
Ag_3SBr		
$Ag_4HgSe_2I_2$		

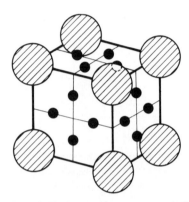

Figure 1.4 The structure of α-AgI. The larger spheres represent iodine atoms. The 12 tetrahedral sites, which are statistically occupied by silver atoms, are represented by filled circles.

atoms form a body-centred cubic sub-lattice, and as such pose no problems. However, on looking at the crystal structure in more detail we discover that there are quite a number of different possible cation sites available: six with octahedral geometry, 12 with tetrahedral geometry and 24 with trigonal geometry. The unit cell shown in Figure 1.4 contains only two AgI molecules and so two Ag^+ ions must therefore be distributed in some way among the 42 possible sites. In the low-temperature polymorph, β-AgI, the silver atoms are exclusively in tetrahedral positions. Hence it would be reasonable to suppose that two of the tetrahedral positions will be 'normal' sites, permanently occupied by atoms, with the others available as interstitial positions for any Frenkel defects formed.

Unfortunately, this simple picture does not hold. Early structure determinations indicated that the ions were randomly distributed among all 42 available sites. This led to the concept of a *molten sub-lattice* of Ag^+ ions moving like a liquid through a fixed matrix of I^- ions. In recent years, a more precise picture of the structure has emerged in which the silver atoms seem to be distributed statistically in the 12 tetrahedral positions. The atoms are unable to distinguish between sites which should be permanently occupied and those which should normally be empty. If we imagine that the atoms are rapidly jumping from site to site we have a very dynamic picture of large numbers of Frenkel defects constantly forming and being annihilated.

Structural studies have shown that the same thing happens in the other materials listed in Table 1.4, and all seem to possess very mobile metal atom arrays. To link this with the earlier parts of this chapter we need to return to the statistical analyses given then. Fortunately, because AgI has the same MX composition that we chose for our examples, we can utilize equation (1.39) to investigate the number of Frenkel defects present in a crystal of the AgI type. This shows that a high number of defects will exist if the energy of formation of the defects is very small and if the temperature is high. These are conditions which apply to the liquid state and so the molten sub-lattice concept is a reasonable one. Our simple theory will not, however, explain why only the metal atoms are affected in this way or why these compounds are rare rather than commonplace.

1.7 Supplementary reading

Frenkel and Schottky defect equilibrium is treated in a number of textbooks, among which are: W.D. Kingery, H.K. Bowen and D.R. Uhlmann, *Introduction to Ceramics*, 2nd edn., Wiley-Interscience, New York (1976); N.N. Greenwood, *Ionic Crystals, Lattice Defects and Non-Stoichiometry*, Butterworth, London (1968); R.A. Swalin, *Thermodynamics of Solids*, 2nd edn., Wiley-Interscience, New York (1972).

A very clear account, together with self-assessment questions, is given in R.F. Davies, *J.Ed. Mod. Mat. Sci. Eng.* **2** (1980) 837.

An excellent review article, which also covers material relevant to Chapters 2 and 3, is J. Corish and P.W.M. Jacobs, in *Surface and Defect Properties of Solids*, Vol. 2, eds. M.W. Roberts and J.M. Thomas, The Chemical Society, London (1975).

The photographic process is well documented, and advertising literature often contains much useful information. A detailed review, together with many literature references, is given by F.C. Brown, in *Treatise on Solid State Chemistry*, Vol. 4, *Reactivity of Solids*, ed. N.B. Hannay, Plenum, New York (1976).

Brief but clear accounts of the photographic process and photochromic glasses are given by K. Nassau, in *The Physics and Chemistry of Colour*, Wiley-Interscience, New York (1983).

Detailed information on photochromic and other glass, together with a comprehensive bibliography is given in D.C. Boyd and D.A. Thompson, 'Glass', in *Encyclopedia of Chemical Technology*, Vol. 11, 3rd edn. Wiley, New York (1980), 807–880.

AgI and related solid electrolytes are reviewed by K. Funke, in *Progress in Solid State Chemistry*, Vol. 11, eds J.O. McCaldin and G. Somorjai, Pergamon, Oxford (1976), 345–402.

2 Atomic mobility: diffusion

2.1 Introduction

One of the most significant aspects of point defects in solids is the way in which they facilitate the movement of atoms or ions through the structure. In this and the following chapter we focus attention upon this role. The ensuing mass transport can be driven by a concentration gradient, in which case it is called *diffusion*, or by a gradient in electrical potential, in which case it is called *ionic conduction*. In this chapter we turn our attention to the phenomenon of diffusion.

In solids, atomic movement can be through the crystalline lattice, and it is this process that we will principally be concerned with. It is called *volume, lattice* or *bulk* diffusion. A little thought will show that point defects are needed for this process to occur at all. However, atoms can also diffuse along surfaces, grain boundaries, dislocations or other macroscopic defects in the crystal. As the regular crystal geometry is disrupted in these regions, atom movement is often much faster than for volume diffusion. Diffusion by way of these pathways is usually referred to as *short-circuit* diffusion.

The speed at which atoms or ions move through a solid is usually expressed in terms of a *diffusion coefficient*, which has units of $m \, s^{-1}$. Not surprisingly, the diffusion coefficient of an atom will vary widely, depending upon external circumstances. Temperature will certainly affect the speed of movement of atoms, as will the geometry of the crystal structure that the atom is moving through. More than this, diffusion coefficients will depend upon whether a chemical reaction accompanies the atom movement, the number of defects present, and so on. Hence, in the literature one will find mention of tracer diffusion coefficients, self-diffusion coefficients, chemical diffusion coefficients and other expressions, all of which derive from the experimental conditions under which the diffusion coefficients were determined. To clarify matters, we list the various diffusion coefficients discussed in this chapter in Table 2.1. The terms themselves will be explained at appropriate points in the following text.

The plan adopted here in treating diffusion has been to break the topic into two parts. In this chapter we discuss the experimental determination of diffusion coefficients. This will allow us to clarify and define the various terms just referred to. Although we take for granted that atoms move through the

Table 2.1 Symbols and terms for diffusion coefficients

Symbol	Meaning	Applicability
D	Self-diffusion coefficient	Random diffusion processes in the absence of a concentration gradient
D^*	Tracer diffusion coefficient	Diffusion when concentration gradients are small
D	Chemical diffusion coefficient	Diffusion in a concentration gradient
D_{AB}	Chemical diffusion coefficient	Total chemical diffusion coefficient for the reaction between A and B
D_A	(Chemical) inter-diffusion coefficient	Diffusion coefficient of one component (A) in a chemical diffusion process

solid by one mechanism or another, such mechanistic considerations are put to one side. The following chapter is then reserved exclusively for interpretation of the experimental results in terms of atom movements and it is here that the importance of point defect populations will be fully appreciated.

We will begin this chapter by considering the simplest case, diffusion which takes place when we have virtually no concentration gradients present or chemical reactions taking place. Firstly let us look at how, experimentally, we determine diffusion coefficients under these conditions.

2.2 Self-diffusion and tracer diffusion

2.2.1 *Experimental determination of self-diffusion coefficients*

When component atoms of a pure crystal diffuse under no chemical or other gradient, the process is that of *self-diffusion*. In such case the atomic movements are likely to be more or less random in nature, the constraints to movement being due to the geometry of the crystal structure. The driving force for the diffusion will be the change in the entropy of the system, which will generally be rather small. Under these conditions, the relevant diffusion coefficient is called the *self-diffusion coefficient* and is given the symbol D.

It is by no means easy, strictly speaking, to measure the self-diffusion coefficient of an atom. However, it is possible to measure something which is a very good approximation to the self-diffusion coefficient, the *tracer diffusion coefficient*, D^*. A typical experimental procedure would be to coat one face of a single crystal of the phase of interest with a thin layer of the same material which we can somehow differentiate from the original crystal. This layer contains the tracer atoms. The coated crystal, called a *diffusion couple*, is then heated at a constant temperature for a known period of time. After this

(a)   (b)

Figure 2.1 Reaction couple used for diffusion experiments. The slabs are carefully polished and oriented slices of MgO. The central dark strip in (*a*) represents a very thin layer of MgO containing radioactive Mg atoms. After the diffusion experiment the radioactive Mg has moved away from the original plane, as in (*b*).

treatment the couple is analysed to determine how far the tracer atoms have moved. In this experiment there *will* be a concentration gradient, and the diffusion coefficient so measured is properly called the tracer diffusion coefficient. However, if layer of tracer atoms is very thin, the concentration gradient will be small, or will rapidly become so, and in these circumstances D^*, the tracer diffusion coefficient, will be a reasonable approximation to the self-diffusion coefficient, D.

As an example, to measure the tracer diffusion coefficient of Mg in MgO, a thin layer of radioactive Mg can be evaporated on to the surface of a carefully polished single crystal of MgO. This can be easily oxidized to MgO by exposing the layers to oxygen gas, after which another carefully polished single crystal slice of MgO is placed on top to form a diffusion couple, as shown in Figure 2.1.

The crystal sandwich is now heated for a known time at a fixed temperature. The whole slab is then carefully sectioned parallel to the original interface containing the radioactive MgO layer, and the radioactivity of each slice, i.e. the concentration of Mg in each section, is determined. A graph of concentration of the radioactive component is then plotted against the

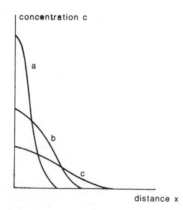

Figure 2.2 Schematic illustration of typical diffusion profiles, which are plots of the concentration of the diffusing species, *c*, against distance, *x*, from the original interface. The three curves (*a*), (*b*) and (*c*) refer to three different heating times, (*a*) being the shortest and (*c*) the longest. The curves are often found to be bell-shaped, as drawn here.

distance from the interface to give a *diffusion profile*, or *concentration profile*. The typical form of such profiles is shown in Figure 2.2.

In order to obtain the diffusion coefficient from such profiles, we make use of some equations which are known as Fick's laws. These were originally set up to describe heat flow, but in general can be applied to any diffusion problem including the one we are considering here. For the present purpose the most useful of these is Fick's second law which relates the change in concentration of the diffusing species with time to the diffusion coefficient. For diffusion along the x direction, the equation is

$$\frac{dc}{dt} = \frac{d}{dx}\left[D* \frac{dc}{dx} \right] \tag{2.1}$$

If the diffusion coefficient $D*$ is assumed to be independent of concentration, the equation becomes a little simpler, viz.:

$$\frac{dc}{dt} = D* \frac{d^2 c}{dx^2} \tag{2.2}$$

where c is the concentration of the diffusing species at position x in the crystal after time t has elapsed.

The equations (2.1) and (2.2) are not particular to the arrangement employed in our experiment and can be solved for any alternative experimental situation employed in practice. In the case described above with the slab geometry illustrated in Figure 2.1, the solution is found to be

$$c = \frac{c_0}{2(\pi D* t)^{1/2}} \exp\left[\frac{-x^2}{4D* t} \right] \tag{2.3}$$

where c is the concentration of the diffusing species at a distance of x from the original interface after time t. $D*$ is the tracer diffusion coefficient and c_0 is the initial concentration on the surface. If we do not employ a sandwich, as in Figure 2.1, but simply an uncovered evaporated layer of Mg on the open surface of the lower crystal, the solution to the equation is

$$c = \frac{c_0}{(\pi D* t)^{1/2}} \exp\left[\frac{-x^2}{4D* t} \right] \tag{2.4}$$

A value for $D*$ can be obtained from equations (2.3) or (2.4) by plotting $\ln c$ versus x. To show why this is so, take logarithms of both sides of equation (2.3), which yields equation (2.5):

$$\ln c = \ln\left[\frac{c_0}{2(\pi D* t)^{1/2}} \right] - \frac{x^2}{4D* t} \tag{2.5}$$

This has the form

$$\ln c = \text{constant} - \frac{x^2}{4D* t} \tag{2.6}$$

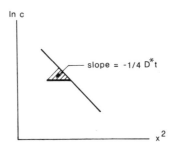

Figure 2.3 A straight line graph of $\ln c$ versus x^2, the slope of which can be used to determine the numerical value of the tracer diffusion coefficient, D^*, from equation (2.6).

Equation (2.4) will, of course, yield an equation of the same form, but with a different constant term.

A schematic graph of equation (2.6) is shown in Figure 2.3, which also indicates that the slope of the graph is equal to $-1/4D^*t$. A measurement of the slope thus allows one to determine the diffusion coefficient at the temperature at which the couple was originally heated.

Remember that equation (2.3) applies only to the diffusion couple geometry shown in Figure 2.1. Hence the slope of the $\ln c$ versus x^2 graph will only be equal to $-1/4D^*t$ for this arrangement of the crystals. Different experimental arrangements or conditions will result in different solutions to equation (2.2). Fortunately the appropriate solutions for most geometrical and chemical situations of interest have long been derived and are to be found in specialist books on diffusion.

It will be clear that the experimental procedure must be carried out with care. The crystals must be carefully polished and cleaned before the experiment, and the sectioning after heating must be exactly parallel to the interface between the slabs in order to obtain true values of the concentration of the radioactive species. Moreover, this experiment has given only one value for the diffusion coefficient, that relevant to the temperature chosen for the heating. If we need the diffusion coefficient over a variety of temperatures, as is usually the case, the experiments must be repeated under these new conditions.

2.2.2 Temperature variation of diffusion coefficients

Both tracer and self-diffusion coefficients are usually found to vary considerably with temperature. This variation can often be expressed in terms of the equation

$$D = D_0 \exp(-E/RT) \tag{2.7}$$

In this equation E is called the *activation energy* of diffusion and is regarded as the energy to move the diffusing species through the solid, R is the gas constant, T is the temperature at which the value of D was measured, and D_0 is a constant term.

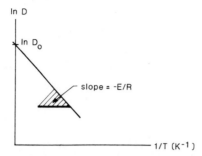

Figure 2.4 An 'Arrhenius plot' of $\ln D$ versus $1/T$, used to determine the activation energy for a diffusion process. The intercept of the line at $1/T = 0$ yields a value for $\ln D_0$ and the gradient yields a value for the activation energy of diffusion.

Table 2.2 Some values for self-diffusion coefficients*

Atom	Matrix	Activation energy (kJ mol^{-1})
Ag	Ag	155
Cu	Ag	193
Fe	Fe	239
C	Fe	85
Ge	Ge	287
P	Ge	45
Li	Ge	239
Mg	MgO	326
O	MgO	258
Be	MgO	152
Ni	MgO	200
Ca	CaO	116
Ti	TiO_2	253
O	TiO_2	247
Li	TiO_2	314
Fe	Fe_3O_4	347
Fe	Fe_2O_3	413
O	Fe_2O_3	330
Co	CoO	159
O	CoO	392
Ni	NiO	251
O	NiO	237
Cr	Cr_2O_3	252
O	Cr_2O_3	416
O	ZrO_2	117
Zr	$CaO.ZrO_2$	382
O	$CaO.ZrO_2$	129
U	UO_2	413
O	UO_2	123

*Note: Literature values for self-diffusion coefficients vary widely, indicating the difficulty of making reliable measurements. The values here are meant to be representative only.

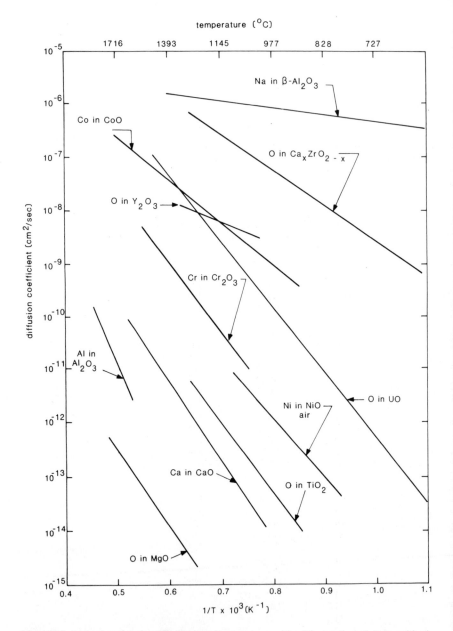

Figure 2.5 Arrhenius plots for diffusion in some common oxides. The y-axis scale is logarithmic, so that values of *D* are plotted directly against 1/*T*. The slope of these graphs yields the activation energy of diffusion as shown in Figure 2.4.

By taking logarithms of both sides of this equation we find

$$\ln D = \ln D_0 - E/RT \qquad (2.8)$$

If a graph of $\ln D$ versus $1/T$ is drawn, the activation energy can be determined from the slope, as the schematic example shown in Figure 2.4 reveals. Such graphs are frequently known as Arrhenius plots, named after the Swedish chemist Svante Arrhenius, who first investigated the activation energies of chemical reactions. Some experimental data, plotted in this way, are presented in Figure 2.5.

Some numerical values of diffusion coefficients will be found in Table 2.2. Others can be calculated from the slopes of the lines in Figure 2.5.

2.2.3 The effect of impurities

Before closing this section we should point out that not all Arrhenius plots are as straightforward as those shown in Figures 2.4 and 2.5. A form of the graph which is frequently found seems to have two straight-line parts to it but with differing slopes, as shown in Figure 2.6.

The point where the two lines interact is called a 'knee'. It is usually found that this point varies from crystal to crystal. The region at low temperatures, to the right of the knee, has a smaller activation energy than the region to the left, which normally corresponds to the high-temperature regime. The lower-temperature region is associated with the impurity content in the crystal, which accounts for its variable interaction with the high-temperature part of the graph. In Figure 2.6, for instance, crystal 1 would have a higher impurity concentration than crystal 2. The two parts of the graph are also known as the *intrinsic* region and the *impurity* region respectively.

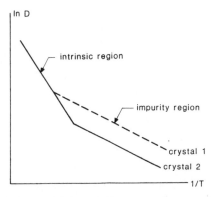

Figure 2.6 A frequently encountered form of Arrhenius plot. The region at higher temperatures is called the *intrinsic* region, and does not vary greatly from crystal to crystal. The lower temperature curves can occur in a variety of positions, dependent upon the impurity content of the crystals. It is thus referred to as the *impurity* region. In the example shown, crystal 1 would have a higher impurity content than crystal 2.

2.3 Chemical diffusion

2.3.1 *Chemical diffusion coefficients*

In the previous section we discussed diffusion in the case where changes in concentration of the diffusing species were unimportant. In practice this is rarely true, and diffusion is usually driven by concentration gradients, or, more precisely, by changes in chemical potential. Under these circumstances the diffusion coefficient cannot be regarded as independent of concentration. The diffusion coefficient measured in these conditions is called the *chemical diffusion coefficient*, and is written $\tilde{D}$.

In the case of chemical diffusion the driving force for the diffusion will be quite different from one system to another, and different for each of the elements in the reaction. Because of the variability of the values obtained experimentally for $\tilde{D}$, it is helpful to try to relate them to self-diffusion coefficients or tracer diffusion coefficients. This objective will form the basis for much of the following discussion.

A generalized formulation of the equations of chemical diffusion is quite complex and will depend upon crystal symmetry and the direction of atom movement, as in the case of tracer diffusion, as well as upon the concentration effects that we consider here. Fortunately, for diffusion along only one dimension, the chemical diffusion coefficient can still be determined by solution of Fick's second law, which is now of the form

$$\frac{dc}{dt} = \frac{d}{dc}\left[\tilde{D}\frac{dc}{dx}\right] \tag{2.9}$$

The problem is how to obtain a value of $\tilde{D}$ using experimentally determined penetration or concentration curves of the sort shown in Figure 2.2 for this new and more complex situation.

One approach is to try to solve the equation analytically for certain rather restricted experimental arrangements. In general, reasonably simple experimental conditions hold when one pure metallic element diffuses into another pure metallic element. Because of this we naturally find that theoretical discussion is often centred around the interdiffusion of two chemically similar metals to form an alloy phase. In order to give a feeling for how diffusion coefficients in chemically reacting systems differ from tracer and self-diffusion coefficients, and how the determination of such diffusion coefficients can be approached in practice, we will stay close to such simple systems in this and the following sections. These provide complexity enough! Rigorous derivations of most of the equations that we come to are not given, as these would tend to defeat the purpose of the presentation. For interested readers, they are to be found in many standard texts concerned with either diffusion or metallurgy, some of which are listed in the Supplementary Reading section at the end of this chapter.

2.3.2 Determination of chemical diffusion coefficients when concentration dependence is weak

Although chemical diffusion coefficients, by definition, are concentration-dependent, it found that this dependence is sometimes fairly weak. This is so for situations where the diffusing atoms do not interact strongly with the surrounding crystal lattice, such as the penetration of small atoms into a relatively open crystal structure by way of an interstitial pathway. Under these circumstances we can approximate equation (2.9) to equation (2.10):

$$\frac{dc}{dt} = \tilde{D}\frac{d^2c}{dx^2} \tag{2.10}$$

and proceed to obtain an analytic solution compatible with the experimental arrangements employed.

As an example of this method we can consider the diffusion of a gas, say nitrogen, into a slab of metal. If the gas pressure is held constant during the process, the concentration of gas at the metal surface will also be constant throughout the duration of the reaction. This condition allows us to solve equation (2.10) analytically. The solution is found to be

$$c(x, t) = c_0[1 - \text{erf}(x/2\tilde{D}t)^{1/2}] \tag{2.11}$$

where c_0 is the constant gas concentration at the surface of the sample, $c(x, t)$ is the concentration of the diffusing species at a distance x from the surface after reaction time t, and $\tilde{D}$ is the chemical diffusion coefficient. The function $\text{erf}(x)$ is given by

$$\text{erf}(x) = \frac{2}{\pi^{1/2}} \int_0^x e^{-x^2}\, dx \tag{2.12}$$

and is called the error function. This is normally tabulated in books of mathematical tables.

Experimentally the situation is now rather similar to the situation that we described for the determination of tracer diffusion coefficients. The penetration curves of the nitrogen gas into the metal must be determined for various temperatures of interest and the values of c, x, and t inserted into equation (2.11) to obtain the corresponding values of D. We can see, therefore, that in these favourable circumstances, a value of the chemical diffusion coefficient is relatively easily found.

2.3.3 The Matano–Boltzmann relationship

Despite the success noted in the preceding section, in general the form of Fick's second law quoted in equation (2.9) cannot be solved analytically, as $\tilde{D}$ depends quite strongly upon c. To obtain values for the diffusion coefficients we must therefore turn to other techniques. One procedure is to use a graphical method which was originally proposed by Matano.

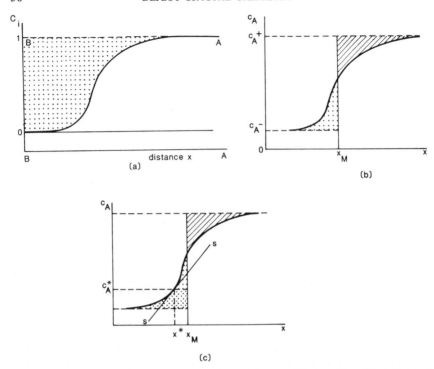

Figure 2.7(*a*) The penetration curve for component A after interdiffusion of A and B. As the total metal content is constant the curve for component B will be the mirror image of that for A. (*b*) The concentration profile of A redrawn to show the position of the Matano plane, at x_M, which is drawn so that the shaded areas are equal. (*c*) The area to be measured, shaded with dots, and the slope of the curve needed, line *s–s*, for the evaluation of the chemical diffusion coefficient at concentration c_A^*.

To illustrate this, we will consider a case where the diffusion couple consists of two metals, A and B, say Cu and Au, and suppose that the concentration profile after interdiffusion has been measured using an electron microprobe or similar technique, to give the result shown in Figure 2.7(*a*). In this figure c_i represents the concentration of component *i*, where *i* is A or B, of course, and the *x* axis gives the distance along the diffusion couple. In our example we see that at the far left we have the pure B and at the far right pure A. The curve shown gives the concentration of component A as we traverse the couple. As the total metal content A + B is constant, the curve which represents the concentration of component B is the mirror image of the curve drawn for A.

In Figure 2.7(*b*) the distribution curve for component A has been redrawn, and so as to make the discussion that follows as general as possible we will call the concentrations of A at the extreme left of the diagram c_A^-, and at the far right c_A^+, rather than 0% and 100%. This will prove of use if A and B are present in alloys, say, instead of being present as pure metals. Figure 2.7 (*b*) also shows that the curve has been divided by a line, at position x_M, to yield two areas which are shown shaded. In the real diffusion couple this line will correspond to a plane normal to the x direction. When this plane is chosen so as to make the two shaded areas on Figure 2.7 (b) equal to one another the plane is called the *Matano plane*.

This definition means that the Matano plane can always be determined from our experimental data. The Matano plane, therefore, makes a convenient reference point for subsequent discussions of atomic movements. It might be thought that this is a rather unnecessary complication. Why not, for instance, use the original interface between A and B before the diffusion experiment to act as our reference point? This is a good idea, but unfortunately, as we shall see later, it is not always easy to locate this plane after diffusion has occurred. In contrast to this, the Matano plane can be located no matter how hard it is to find the original position of the interface.

Once having drawn the Matano plane we can now determine the chemical diffusion coefficient, $\tilde{D}$, at any value c_A^* that we care to choose, as it is simply given by

$$\tilde{D} = - \text{[area under curve between } c_A \text{ and } c^*]/[\text{slope of the curve at } c_A^*]$$
(2.13)

Referring to Figure 2.7(c), we see that the area we need is shaded, and the slope that we need is that of the line labelled $s-s$. By repeatedly changing the value of c_A^*, and recalculating the new areas and slopes, we can determine how $\tilde{D}$ varies with changing concentration, c_A, and with distance from the Matano interface, x^*. If $\tilde{D}$ is the same on both sides of the Matano plane then the curve will be symmetrical with respect to this interface. In Figures 2.7(b) and (c) this is not so and the curves have been drawn unsymmetrical on purpose.

The diffusion coefficient that is found from the Matano–Boltzmann analysis is the diffusion coefficient that describes the reaction between A and B and not the diffusion coefficients for either of the components separately. For this reason it is sometimes useful to write it as $\tilde{D}_{AB}$. We will try to gain some idea of the way in which the diffusion coefficients of A and B separately contribute to $\tilde{D}_{AB}$ later.

It is by no means intuitively obvious that the result given in equation (2.13) is correct. Approximations or assumptions are often made in deriving equations of general validity, and it is important to be aware of these limitations should they exist. Thus, it is necessary to gain some idea of the way in which equation (2.13) was arrived at. In so doing we will uncover a number of other pieces of useful information.

In order to make a start we first write equation (2.9) in a form relevant to the results shown in Figure 2.7(a). This is

$$\frac{dc_i}{dt} = \frac{d}{dx} \tilde{D} \frac{dc_i}{dx}$$
(2.14)

where c_i is the concentration of component i at position x in the sample after a reaction time t and $\tilde{D}$ is the relevant chemical diffusion coefficient. In the our case i is either A or B. The most important assumption of the analysis which follows is that c_i is a unique function of the position at which it is measured, x, and the time over which the diffusion has occurred, t. If the function is written λ, then the assumption made can be written algebrically as

$$\lambda = x/\sqrt{t} \qquad (2.15)$$

This apparent sidestep can be used to transform equation (2.14) into a new form that will prove of more use. First of all, let us replace the left-hand side of equation (2.14), dc_i/dt, by an alternative expression arrived at by differentiating equation (2.15) with respect to t, thus:

$$\frac{d\lambda}{dt} = -1/2(xt)^{-3/2}$$

$$\frac{dc_i}{dt} = -1/2(xt)^{-3/2}\frac{dc_i}{d\lambda}$$

Hence

$$\frac{dc_i}{dt} = \frac{-\lambda}{2t}\frac{dc_i}{d\lambda} \qquad (2.16)$$

In a similar way, by differentiating equation (2.15) with respect to x, we can replace the right-hand side of equation (2.14), thus:

$$\frac{d\lambda}{dx} = t^{-1/2}$$

$$\frac{dc_i}{dx} = t^{-1/2}\frac{dc_i}{d\lambda}$$

$$\frac{d^2c_i}{dx^2} = t^{-1}\frac{d^2c_i}{d\lambda^2} \qquad (2.17)$$

When this is done we find

$$\frac{-\lambda}{2}\frac{dc_i}{d\lambda} = \frac{d}{d\lambda}\left[\tilde{D}\frac{dc_i}{d\lambda}\right] \qquad (2.18)$$

We can now redraw Figure 2.7 using the function λ to replace x. The shaded area shown in Figure 2.7(c) is then given by the integral from c_A^- to c_A^* of equation (2.18). The limits of the curve, originally chosen as $x = -\infty$ to $x = +\infty$, corresponding to c_A^- and c_A^+, are equivalent to $\lambda = -\infty$ and $\lambda = +\infty$. On integrating equation (2.18), we can write

$$-\tfrac{1}{2}\int_{c_A^-}^{c_A^*}\lambda\,dc_i = \tilde{D}\left[\left(\frac{dc_i}{d\lambda}\right)_{c_i=c_A^*} - \left(\frac{dc_i}{d\lambda}\right)_{c_i=c_A^-}\right] \qquad (2.19)$$

Now we know that at $c_i = c_A^-$, i.e. at $x = -\infty$, the rate of change of c_i with x is zero, as we have pure A at that point, that is

$$\left(\frac{dc_i}{d\lambda}\right)_{c_i = c_A^-} = 0 \tag{2.20}$$

hence, the chemical diffusion coefficient at c_A^*, $\tilde{D}_{c_A^*}$, is given by

$$\tilde{D}_{c_A^*} = -\frac{1}{2} \int_{c_i^-}^{c_A^*} \lambda\, dc_i \bigg/ \left(\frac{dc_i}{d\lambda}\right)_{c_A^*} \tag{2.21}$$

where $(dc_i/d\lambda)_{c_A^*}$ is the differential of the concentration with respect to λ at c_A^*. It now helps to reconvert λ to $x/\sqrt{t}$, and after this substitution we can write

$$\tilde{D}_{c_A^*} = \frac{1}{2} \int_{c_i^-}^{c_A^*} x t^{1/2}\, dc_i \bigg/ t^{1/2}\left(\frac{dc_i}{dx}\right) \tag{2.22}$$

$$\tilde{D}_{c_A^*} = -\int_{c_i^-}^{c_A^*} x\, dc_i \bigg/ 2t\left(\frac{dc_i}{dx}\right)_{c_i^*} \tag{2.23}$$

The last two equations, (2.22) and (2.23), which are equivalent to one another, are each known as the Matano–Boltzmann equation. Although they look extremely unpleasant, we know that they are not too formidable in practice, as they simply express in mathematical terms the verbal equation (2.13). The integrals forming the numerators of equations (2.22) and (2.23) simply represent the shaded area in Figure 2.7(c) and the denominators in these equations represent the slope of the curve at c_A^*.

It is important to remember that we did make the assumption that c_i was a unique function of position in carrying out the analysis. This may not always be so, but it does seem to hold reasonably often. In such cases another useful piece of information which relates the Matano plane to the extent of diffusion can be derived. This comes about in the following way. Any particular value of concentration, at the temperature in question, is directly tied to one value of λ, which is one value of $x/\sqrt{t}$. As the time, t, increases, the value of x increases, and so we can imagine that the plane x corresponding to the concentration c_i moves through the diffusion couple at a steady rate during the course of the diffusion reaction. Its distance from the Matano plane, x, at any time t is given by

$$x^2 = kt \tag{2.24}$$

The factor k is simply a proportionality constant which is known as the *penetration constant*. As we shall see in the following section, equation (2.24) can be used to gauge the separate magnitudes of the diffusion coefficients of the components A and B themselves.

2.4 Chemical diffusion, intrinsic diffusion and self-diffusion

2.4.1 *The Kirkendall effect*

In the preceding section we used the Matano interface as a reference plane when it was necessary to measure distances in a diffusion couple. Intuitively it seems far simpler to use the initial interface between the two reactants, A and B, for this purpose. The problem experimentally is how to locate this plane after reaction has occurred.

Let us think about this a little. If we imagine our diffusion couple after heat treatment, the central region will consist of the interdiffusion phase. If we suppose that A diffuses twice as quickly across the interface into B as B does in the opposite direction, then clearly the amount of alloy phase on one side of the interface will be twice as much as on the other side, and the diffusion coefficient, or at least the relative diffusion coefficients, can easily be determined by simple measurement. So why not mark the initial interface? We could do this by, for example, placing a few inert markers, platinum wires, say, at the interface before the heating cycle is started. The fact of the matter is that the inert markers will not always remain in place at the interface.

This shift of markers is known as the *Kirkendall effect*. It was first observed in an experiment in which a block of α-brass (70% Cu: 30% Zn) was embedded in a block of copper. The brass was wrapped around with fine Mo wires which were to act as the inert marker, as shown in Figure 2.8(*a*). After heating, it was found that the separation of the wires had decreased. The reason for the marker shift turned out to be fairly clear; the Zn atoms diffused out of the brass block faster than the Cu atoms moved in, and the block appeared to shrink.

We have illustrated this effect schematically in Figures 2.8(*b*) and (*c*) for the system that we are considering, that of the interdiffusion between two components A and B. Assume that markers are placed at the interface between the components, as shown in Figure 2.8(*b*), that the diffusion of A is faster than the diffusion of B and that the volume of the system after the diffusion experiment is the same as before it. That is, alloy formation does not alter the total volume of the couple. After some time, the volume of A which has passed to the right of the marker R is greater than the volume B which has passed to the left of R. This is shown in Figure 2.8(*c*), where we now see that the volume to the left of the marker is smaller than the volume to the right. The opposite happens at the marker L. Clearly, we started out with equal volumes on both sides of the marker, and so it appears, in an experimental observation, that the marker R has moved to the left and marker L has moved to the right. This is so even if we have no overall volume change, as we specified above. The separation between the markers, w, thus appears to decrease.

In reality the extent of the Kirkendall effect is difficult to estimate. There will invariably be some volume change, although this may be small. In addition, if one component leaves one area of sample faster than the other component

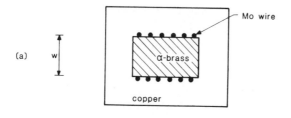

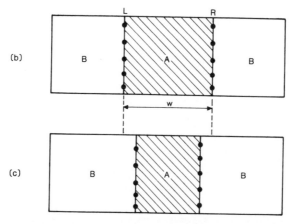

Figure 2.8(*a*) Schematic illustration of the original demonstration of the Kirkendall effect. (*b*) The situation in (*a*) before reaction. (*c*) The situation in (*a*) after reaction, if component A diffuses faster than component B. The separation of the markers, *w*, appears to decrease.

moves in, we are likely to get voids formed. This frequently happens in practice, and has an effect on the actual marker displacement that is measured. It is therefore difficult to treat the Kirkendall effect theoretically in a general way. One situation has been analysed, however, and we briefly discuss this now, as it will allow us to separate the contribution of the diffusion coefficient of component A from that of component B in the reaction.

2.4.2 *Intrinsic diffusion coefficients*

The chemical diffusion coefficient, $\tilde{D}$, for our A–B system is made up of two so-called *intrinsic diffusion coefficients*, $\tilde{D}_A$ and $\tilde{D}_B$. These, of course, are equivalent to the chemical diffusion coefficients of each of the separate components in our reaction. In general $\tilde{D}_A$ and $\tilde{D}_B$, as well as $\tilde{D}$, are concentration-dependent. It is generally not possible to relate these quantities to each other over all the concentration ranges that apply in the diffusion couple, but if we use the moving marker plane as a reference plane some relationships can be found.

This is not simply an arbitrary choice of origin, of course. Suppose, for example, we wish to measure the diffusion of a dye molecule in water. The dye could simply be injected into the water at a certain marked spot, and its spread observed visually. If we have a flowing body of water, such as a river, then we can still use the same technique. However, two factors are now operating. Firstly, the dye is diffusing outwards, as before, but also the whole body of water in which the dye diffuses is being swept along downstream. To study the diffusion process alone it is therefore sensible to take a reference point which moves at the same speed as the flowing water. That is, the point in the river where the dye is injected is taken as the moving reference point, rather than the static point on the river bank from where the dye was put into the following stream. Using the moving Kirkendall marker plane is the analogous situation in our diffusion couple.

The analysis is straightforward, and will not be given in any detail here. The references at the end of the chapter contain full information for those interested. However, it is important to be aware of the assumptions that need to be made in order to arrive at the relationship that we require. Remaining with the simple example of the interdiffusion of components A and B as before, we must make the following assumptions:

(i) A continuous solid solution forms between A and B.
(ii) The molar volumes V_A and V_B are equal and independent of concentration. Consequently the total volume of the couple remains constant and we can write $V_A = V_B = V_M$. In addition, the cross-section of the couple is taken as remaining constant.
(iii) The diffusion couple is infinite, so that the concentrations of A and B at the ends of the couple are constant.
(iv) Diffusion takes place only in the direction perpendicular to the original contact interface between A and B, which we take as the x-direction.

With these assumptions, and remembering that we are restricted to the Kirkendall plane, it can be shown that

$$\tilde{D} = N_A \tilde{D}_B + N_B \tilde{D}_A \qquad (2.25)$$

where N_A is the mole fraction of component A and N_B the mole fraction of component B. With only two components,

$$\left. \begin{array}{l} N_A + N_B = 1 \\ N_A = c_A/C \\ N_B = c_B/C \end{array} \right\} \qquad (2.26)$$

where C is the total concentration, $c_A + c_B$. Equation (2.25) is known as the *Darken equation*.

If the molar volumes are not equal, we find

$$\tilde{D} = N_A V_A \tilde{D}_B + N_B V_B \tilde{D}_A \qquad (2.27)$$

Although the derivation of equations (2.25) and (2.27) is restricted by the assumptions made, the analysis does not depend upon any mechanism chosen for the diffusion, and simply states that, in the moving marker plane, the chemical diffusion coefficient $\tilde{D}$ is simply the sum of the two interdiffusion coefficients, in the proportions defined by equation (2.25) or (2.27).

In the previous section we made the assumption that a plane of constant composition moves through the crystal with a constant velocity. The Kirkendall plane fits this description and so we can suppose that the movement of the Kirkendall plane (measured with respect to the Matano plane, for example) is proportional to $\sqrt{t}$, so that:

$$x_K = a\sqrt{t} \tag{2.28}$$

where x_K is the position of the marker plane, and a is a constant.

The velocity of movement, v_K, is given by:

$$v_K = dx_K/dt = d(a\sqrt{t})/dt = a/2\sqrt{t} = x_K/2t \tag{2.29}$$

The analysis which led to equations (2.25) and (2.27) can be combined with this result to give

$$v_K = \tilde{D}_A - \tilde{D}_B\left(\frac{dN_A}{dx}\right) \tag{2.30}$$

Equations (2.25) and (2.29) at last allow us to reach our goal and determine separate values for both $\tilde{D}_A$ and $\tilde{D}_B$. We make use of the following experimental procedure. Firstly, $\tilde{D}$ is measured in the Kirkendall plane using the Matano–Boltzmann method. This gives $(\tilde{D}_A + \tilde{D}_B)$ from equation (2.25). Then the slope of the curve is measured at the Kirkendall plane, which gives us dN_A/dx. Finally the displacement of the Kirkendall plane from the Matano plane, x_K, is measured and, as the time of the diffusion experiment, t, is known, we can calculate $(\tilde{D}_A - \tilde{D}_B)$. We now have sufficient information to calculate values for both $\tilde{D}_A$ and $\tilde{D}_B$.

We can also use the displacement of the Kirkendall plane, x_K, for the evaluation of activation energies of diffusion, by utilizing the following procedure. Recall that the diffusion coefficient, $\tilde{D}$, can be written as

$$\tilde{D} = \tilde{D}_0 \exp(-E/RT) \tag{2.31}$$

where for chemical diffusion $\tilde{D}$, $\tilde{D}_0$ and E are all dependent upon the concentration of the diffusing species. However, $x^2_K/\tilde{D}t$ is a dimensionless quantity, and if we substitute this into equation (2.31) we find

$$x_K = (K\tilde{D}_0t)^{1/2} \exp(-E/RT) \tag{2.32}$$

where K is equal to $x^2_K/\tilde{D}t$. The value of the pre-exponential constant need not concern us, but if we plot $\ln x_K$ versus $1/T$, the reciprocal of the diffusion temperature, we can directly obtain the activation energy for the chemical diffusion process in the Kirkendall plane.

2.4.3 *The relationship between chemical diffusion and self-diffusion*

At the outset of this chapter we stated that Fick's laws could be applied to the problems of atomic diffusion, and Fick's second law, equation (2.1), was taken as the starting point for discussions of self-diffusion. For many purposes it is more convenient to start from a slightly different equation, Fick's first law, which we can write as

$$J_i = D_i \frac{dc_i}{dx} \tag{2.33}$$

where J_i is the flow of atoms of type i diffusing in the solid, D_i is the diffusion coefficient of atoms i, and c_i their concentration at position x. This equation implies that equilibrium will be achieved, and atom flow will cease, when the concentration gradient dc_i/dx equals zero.

Now thermodyanamics makes it clear that the system will be in equilibrium when the chemical potential, rather than the concentration, of each component is uniform in the system. This suggests that a more correct form of Fick's first law should be

$$J_i = -L_i \frac{d\mu_i}{dx} \tag{2.34}$$

or, alternatively

$$J_i = -L_i \frac{d\mu_i}{dc_i} \frac{dc_i}{dx} \tag{2.35}$$

where μ_i is the chemical potential of component i, and L_i is called a *phenomenological coefficient*. The flow of atoms, J_i, is now proportional to the gradient of chemical potential, rather than concentration. As chemical potential is difficult to estimate and concentrations are relatively simple to determine experimentally, the form of Fick's first law given in equation (2.33) is usually preferred in experimental work. However, the two equations above, (2.34) and (2.35), do allow us to relate intrinsic diffusion coefficients to tracer diffusion coefficients. This valuable link forms the subject of the following discussion.

As before, it is difficult to obtain an analytical solution to the problem and a number of assumptions are needed before even an approximate solution can be derived. However, if we return to our example of interdiffusion of two metals A and B, and we assume that these metals do not chemically or physically interact, we can obtain a relationship between the intrinsic diffusion coefficient and the coefficient L which will be valid in the Kirkendall plane.

The derivation, found in the works cited in section 2.7 (Supplementary Reading), shows that, for component A,

$$\tilde{D}_A = \frac{L_A RT}{c_A} \left[1 + \frac{\partial \ln \gamma_A}{\partial \ln N_A} \right] \tag{2.36}$$

where $\tilde{D}_A$ is the intrinsic diffusion coefficient of component A, c_A its concentration, N_A its mole fraction and γ_A its activity coefficient, all in the Kirkendall plane. The symbols R and T represent the gas constant and the absolute temperature respectively. The expression in square brackets is called the *thermodynamic coefficient*.

Now this equation on its own is rather cumbersome, and not of direct use. However, if we recall the process of tracer diffusion, the tracer diffusion coefficient D^* of metal A applies when the concentration gradient of component A becomes very small. In such dilute solutions, the activity of A, γ_A, will be unity, and as $\ln 1$ is zero, we can rewrite equation (2.36) in the form

$$\tilde{D}_A = D_A^* = \frac{L_A RT}{c_A} \tag{2.37}$$

The intrinsic diffusion coefficient, $\tilde{D}_A$, is thus related to the tracer diffusion coefficient D_A^* by the equation

$$\tilde{D}_A = D_A^* \left[1 + \frac{\partial \ln \gamma_A}{\partial \ln N_A} \right] \tag{2.38}$$

We see then that the intrinsic diffusion coefficient can be regarded as the tracer diffusion coefficient multiplied by a factor, the thermodynamic coefficient, which accounts for concentration changes and the concurrent changes in thermodynamic potential.

If we recall equation (2.25),

$$\tilde{D} = N_A \tilde{D}_B + N_B \tilde{D}_A \tag{2.25}$$

and if we take

$$\partial \ln \gamma_A / \partial \ln N_A = \partial \ln \gamma_B / \partial \ln N_B \tag{2.39}$$

which is a standard equation of thermodynamics, we can write

$$\tilde{D} = (N_A D_B^* + N_B D_A^*) \left[1 + \frac{\partial \ln \gamma_A}{\partial \ln N_A} \right] \tag{2.40}$$

by substituting for $\tilde{D}_A$ and $\tilde{D}_B$ in equation (2.25) from equations of the type given by (2.38), for both A and B. Moreover, from equations (2.38) and (2.39), we find

$$\frac{\tilde{D}_A}{\tilde{D}_B} = \frac{D_A^*}{D_B^*} \tag{2.41}$$

These relationships between the chemical diffusion coefficient, the intrinsic diffusion coefficient and the tracer diffusion coefficient are called the *Darken relations*. They contain only values which may be measured experimentally, and so they can be tested in the laboratory. In general they are found to hold well. As tracer diffusion coefficients are similar in magnitude to self-diffusion coefficients, the Darken relations can also be used with a fair degree of confidence with these latter quantities.

Finally, we need to draw attention to the fact that the treatment above, although it is very instructive, is limited in its applicability. Diffusion is not an equilibrium situation, and the use of equilibrium thermodynamics may not be justified. Indeed, diffusion should be considered in terms of irreversible thermodynamics. This treatment is beyond our needs here, but the Supplementary Reading section lists sources where the topic is taken further.

2.4.4 *Diffusion in multiphasic alloy systems*

In the preceding discussion of intrinsic diffusion coefficients and the Kirkendall effect, attention was focused upon single-phase materials. The equations discussed then, together with the Matano–Boltzmann procedure, can also be used in more complex systems containing intermediate phases.

Let us look at the situation in which an intermediate alloy forms in a system with two components A and B, both of which have solid solution ranges. (Solid solutions are defined and treated in more detail in Chapter 4.) The conventional representation of the phase diagram for such a system is shown in Figure 2.9(a). In a diffusion experiment at a temperature of T_D, continued reaction will ultimately produce a state in which the equilibrium situation represented by the line T_D on Figure 2.9 will be reached. This shows that the solid solution α will coincide with solid solution γ on one side of the diffusion couple, while at the other side γ will coincide with β. Far from the original interface we will have pure components A and B. This situation is shown schematically in Figure 2.9(b).

Focusing attention on metal A, the concentration of this component will remain constant until we reach the solid solution region α. The concentration will then fall across the phase until we reach the solid solution limit, c_α. We will then have discontinuity in the concentration of A, which will jump to the value c_γ, which is compatible with the A-rich side of the alloy phase γ, as shown in Figure 2.9(c). The concentration of A then falls away again to a value of c_γ^* at the other side of the phase. Another discontinuity then results, because the concentration of component A in the β solid solution cannot exceed c_β. In the β solid solution phase the concentration of A will fall gradually to zero, and we arrive at pure B.

Even in this more complex situation the various diffusion coefficients can still be determined by using the Matano–Boltzmann analysis and associated equations, provided that we can obtain penetration curves. In reactions with uneven diffusion coefficients, markers at the initial boundary will also move; the distance from the starting position after a time t being given by an equation similar to equation (2.28), which in the present case would be

$$x_{\alpha\beta} = k^{\alpha\beta} t^{1/2} \qquad (2.42)$$

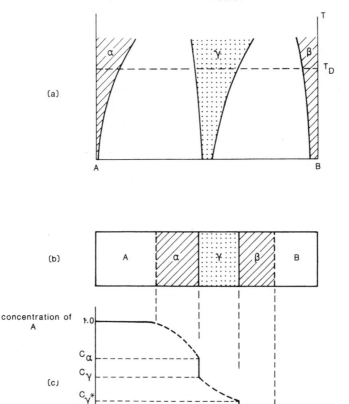

Figure 2.9(*a*) Schematic phase diagram for a two-component system A and B which forms solid solutions α and β and an intermediate phase γ. (*b*) The situation holding after pure A and B have been allowed to react for some time. (*c*) Graph of the concentration of component A across the diffusion couple shown in (*b*).

where $x_{\alpha\beta}$ is the distance that the boundary has moved, and $k^{\alpha\beta}$ is a rate constant. Finally, we should note that, if $\tilde{D}$ can be assumed constant within any of the single-phase regions, we can once again apply analytic solutions to derive the diffusion coefficients.

Exactly the same procedure can be applied to a system in which several intermediate phases occur. We can still obtain the diffusion coefficients that we require provided the phases separate enough, physically, to obtain clean penetration curves. In fact such multiphase diffusion experiments are of some interest, as, in principle, if we wait long enough, all equilibrium phases should be present in sequence along the couple.

2.5 Ambipolar diffusion

2.5.1 *Diffusion in ionic materials*

In previous sections in this chapter we have considered the diffusion process to be driven by changes in concentration and not considered other constraints. This works perfectly well in many cases, particularly for self-diffusion and for the interdiffusion of metal atoms in an alloy. In many cases, though, movement of charged particles, ions or electrons, is of importance. In considering such diffusion, which is referred to as *ambipolar diffusion*, we will see that one other constraint plays an important role. This constraint is that of local charge neutrality, which must be conserved.

The problem of charge neutrality is of greatest importance in diffusion reactions between non-metals. In general, suppose such a reaction is taking place, and that it is maintained by the transport of *ions* rather than atoms through a crystalline material. It is to be expected that these ions would have different mobilities, and that one ion would outstrip another during the reaction. However, any tendency for this to happen would be balanced by an electric charge gradient. This will act in such a way as to increase the mobility of the slower ions and decrease the mobility of the faster ions, so that equality is achieved.

In these cases we would expect treatments of diffusion, which are essentially thermodynamic in origin, such as the Darken relationships mentioned in the previous section, to still hold good, but that the thermodynamic factor would include not only concentration terms (the activity) but also electrochemical terms. This is so, and provided that we use electrochemical potential rather than just chemical potentials, the Darken analysis can be repeated and analogous equations formulated.

In order to examine this concept in a little more detail it is instructive to consider some typical experimental situations. For this purpose we have chosen to focus attention upon the interdiffusion of two ionic compounds to form a solid solution. The principles outlined here will be of more general validity, though, and also apply to instances where solid state reactions are taking place.

2.5.2 *Solid solution formation*

This is conceptually the simplest case to analyse. We assume that the end components can interdiffuse to form a complete solid solution, as in the case of metals A and B, considered earlier. Now, though, the reactants are supposed to be the compounds AX and BX, and reaction involves ionic diffusion. Typical examples would be NiO–MgO, KBr–AgBr and $CoAl_2O_4$–$MgAl_2O_4$. The situation is shown schematically in Figure 2.10.

Let us consider the reaction between NiO and MgO to form a mixed crystal

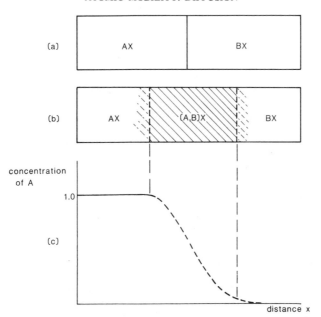

Figure 2.10 Schematic illustration of a diffusion couple in which solid solution forms between the starting phases AX and BX. (*a*) The situation before reaction. (*b*) The situation after reaction. Note that the boundaries between AX and the solid solution and BX and the solid solution will be diffuse. (*c*) Plot of the concentration of component A versus distance across the couple.

of composition $Ni_xMg_{1-x}O$. As the oxygen packing is the same in both NiO and MgO we expect only an interdiffusion of Mg^{2+} and Ni^{2+} ions to take place during reaction. A Kirkendall marker placed at the initial interface should not move under these circumstances. The diffusion profile obtained can be analysed by way of the Matano procedure. An example of a diffusion profile obtained in practice is shown in Figure 2.11.

So far this reaction appears to behave exactly as the alloy formation reaction considered earlier, and the analogy can be continued. The flux of atoms of type i will be given by equations of type (2.34), (2.35) and (2.36), so we can write

$$J_i = \frac{c_i \tilde{D}_i}{RT} \cdot \frac{\mathrm{d}\bar{\mu}_i}{\mathrm{d}x} \tag{2.43}$$

where $\bar{\mu}_i$ is written for the electrochemical potential of the species i. This takes into account the valence of the migrating particles and the electric field in which they move. The situation can now be analysed theoretically to produce Darken equations which we would expect to contain some extra terms compared to those quoted earlier.

The analysis will not be given here, as it turns out that in the NiO–MgO system we also have a small concentration of mobile electrons that arise

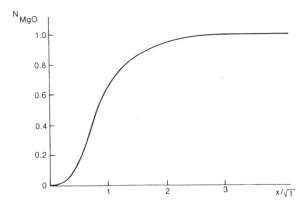

Figure 2.11 An experimentally determined concentration profile for the system NiO–MgO heated at 1370 °C in air. Note that the curve is asymmetrical. Redrawn from S.L. Blank and J.A. Pask, *J. Amer. Ceram. Soc.* **52** (1969) 669.

because of the slightly non-stoichiometric behaviour of NiO, as explained in Chapters 6 and 7. These electrons always allow local charge balance to be maintained, as they migrate very quickly compared to the ions themselves, so that the valence and field effects are cancelled out. In this case the unmodified Darken relations hold, so that we find

$$\tilde{D} = N_{\text{NiO}}\tilde{D}_{\text{Mg}} + N_{\text{MgO}}\tilde{D}_{\text{Ni}} \qquad (2.44)$$

When we consider systems in which reaction is strictly by way of ionic diffusion only, and no mobile electrons are present, we cannot make this simplification. An example is given by the KCl–RbCl system. Here both anion and cation diffusion are significant and markers placed at the original interface will move appreciably during reaction. Because of the absence of electron transport, the chemical diffusion coefficient $\tilde{D}$ will be given by an expression more complex than that of equation (2.44) and will contain an electrical field term in one form or another.

2.5.3 *Tarnishing and oxidation reactions*

Tarnishing and oxidation occur when a metal is attacked by a gaseous atmosphere, often oxygen, to produce a thin layer of product phase. Due to the industrial and economic importance of such reactions they have been very extensively studied. Our purpose here is simply to look at these reactions in the context of the problem of diffusion. Because of this we will consider only the case when a coherent film of oxide forms on the metal surface. In such a case further reaction can proceed only if ions or atoms can diffuse across the film, either from the outside gaseous phase into the inner metal layer, or else from the metal out to meet the gas.

The situation after the initial reaction has taken place is shown in Figure 2.12. A number of ways in which the reaction can proceed can be envisaged. The scheme illustrated in Figure 2.12(a) shows diffusion of metal ions outward from the metal towards the gas atmosphere. To maintain electrical neutrality in the system, this diffusion must be accompanied by a parallel diffusion of electrons. The two processes operating in tandem will result in new film growth at the outer surface of the film. The scheme illustrated in Figure 2.12(b) shows the diffusion of

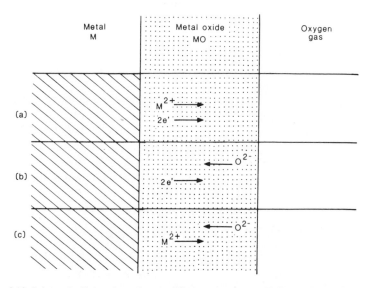

Figure 2.12 Schematic illustration of some diffusion processes which can allow the continued oxidation of a metal M to an oxide MO. (a) Diffusion of cations and electrons allows further reaction to take place at the outer surface of the oxide. (b) Counter-diffusion of anions and electrons allows reaction to take place at the inner surface of the oxide. (c) Counter-diffusion of both anions and cations allows the reaction to take place at both surfaces.

oxygen ions into the film, and a consequent oxidation of the metal and film growth at the inner surface of the film at the metal/metal oxide boundary. This must also involve a movement of electrons from metal to the outer surface in order to maintain charge neutrality. In the third scheme, Figure 2.12(c), we see the only mechanism in which electrons are not needed, the counter-diffusion of cations and anions. Other schemes not shown in this figure can also be written, of course.

In all these reactions, electrons will be the most mobile particles, but due to the necessity of maintaining charge neutrality, their movement may be controlled by the ionic species with the lowest diffusion coefficient. In all cases, the concentration of the metal and oxygen gas will be constant, and it may be possible to consider the chemical diffusion coefficient to be independent of concentration. In such a case, the relevant diffusion equations can be solved analytically, following the procedures outlined in section 2.3.2.

2.6 Conclusions

In this chapter we have discussed diffusion, mainly from the point of view of determination of diffusion coefficients, and usually with respect to isotropic binary systems. Moreover, we have almost always considered diffusion along only one crystallographic direction. The atomistic aspects of diffusion have largely been ignored, and equilibrium thermodynamics has been used to arrive at many of the equations quoted.

From a practical point of view, very many systems are not isotropic, and diffusion at surfaces or along dislocations may well dominate any reactions taking place. In addition, the ease of atomic movement during bulk diffusion will depend upon direction in a majority of crystal structures, and so diffusion is more properly treated in terms of tensor notation. From a theoretical point of view, a discussion of diffusion in terms of irreversible thermodynamics is preferable to a discussion in terms of equilibrium thermodynamics, and allows many of the approximations involved in the latter part of the chapter to be understood. These matters, though, do not involve truly new concepts but principally an expansion of the ideas presented earlier. On the other hand, ideas relating to of the mechanism of diffusion, particular from an atomic point of view, have not yet been covered. It is now time to redress this balance, and Chapter 3 considers diffusion from a mechanistic viewpoint.

2.7 Supplementary reading

There are many books which treat the topic of diffusion in depth. One of the most readable is P.G. Shewmon, *Diffusion in Solids*, McGraw-Hill (1963).

Many experimental results are to be found in P. Kofstad, *Nonstoichiometry, Diffusion and Electrical Conductivity in Binary Metal Oxides*, Wiley-Interscience, New York (1972).

The subject is also treated extensively in textbooks of metallurgy and materials science. A good account is given in W.D. Kingery, H.K. Bowen and D.R. Uhlmann, *Introduction to Ceramics*, 2nd edn., Wiley-Interscience (1976).

A very clear account of diffusion, together with self-test examples, is given in R. Metselaar, *J. Mater. Ed.* **6** (1984) 229; **7** (1985) 653.

Worked examples covering a wide range of diffusion problems are given in the very useful book, R.G. Faulkener, D.J. Fray and R.D. Jones, *Worked Examples in Mass and Heat Transfer in Materials Engineering*, Institution of Metallurgists, London (n.d.).

An advanced review article which summarizes diffusion theory is A.D. LeClaire, Chapter 1 in *Treatise on Solid State Chemistry*, ed. N.B. Hannay, Plenum, New York (1976).

3 The atomic theory of diffusion

3.1 Introduction

In Chapter 2 the process of diffusion was discussed largely from the experimental viewpoint of determination of diffusion coefficients. However, we know that diffusion takes place by atom movement, and it is of some interest to try to relate the experimental data to these more fundamental processes. Of course this has more than an academic interest, for if the atomic movements which constitute diffusion can be understood, then those processes which depend upon diffusion can be controlled more knowledgeably and materials can be chemically and physically tailored more precisely.

Diffusion simply means that atoms are moving through the crystal structure. If we ignore atom movement via dislocations, grain boundaries or surfaces, all of which are important in practice, we are left with a consideration of diffusion through the crystal lattice. Some crystal structures are open, and will contain atoms in sites which allow them great freedom of movement, while others are in more restricted environments. However, it is always found that point defects or larger aggregates of point defects must be present before diffusion can take place. An atom can only jump from a normal site if the site at the end of the displacement is empty, that is, if a vacancy exists. Similarly, if an atom has to be displaced into a normally unoccupied position before it can move, we have created an interstitial type of defect. Thus, we need to build on the ideas contained in Chapter 1, and try to combine them with notions of atom movement to reproduce the experimental results detailed in Chapter 2.

To do this, we will once again start with the process of self-diffusion, and attempt to explain the results of section 2.2 in terms of atomic motion. These ideas will be extended to a consideration of how the imposition of potential barriers can change the overall characteristics of the diffusion process. Although the most general potential to take is that of the electrochemical potential, we will restrict our discussion mainly to the simpler case of an electrical potential, in which chemical potential effects can be largely overlooked. In this way we come inevitably to a consideration of ionic conductivity, a topic of considerable importance with relevance to solid state batteries and other devices.

We shall see that rather simple ideas take us a long way in explaining both diffusion and ionic conductivity and that Schottky and Frenkel defect equilibria have an important role to play in the story.

3.2 Self-diffusion mechanisms

3.2.1 *Atomic migration*

Let us first look at some schematic ways in which we can think of atoms moving through the crystals. For normal crystals this will probably be by way of individual atom jumps from one stable position to another, although in circumstances where very fast diffusion takes place, as in the example of α-AgI described in Chapter 1, other mechanisms may hold. Some ways in which these individual jumps can take place are illustrated schematically in Figure 3.1. Remember that in real crystals the paths will be more complex, and controlled by the three-dimensional geometry of the structure. Crystal structure models should always be consulted if one is intent upon mapping real diffusion paths in solids.

Looking at the pathways shown in Figure 3.1, we can see that movement of a diffusing atom into a vacancy corresponds to movement of a vacancy in the other direction. This process is therefore frequently referred to as *vacancy diffusion*. In practice it is often very convenient, in problems where vacancy diffusion occurs, to ignore atom movement and to focus attention upon the diffusion of the vacancies as if they were real particles, in the same way that electronics makes use of electron holes in describing some electrical transport phenomena. It is likely that this mechanism is of considerable importance in close-packed solids. The diffusion of interstitial atoms by way of the *interstitial diffusion* mechanism is important in the hardening of steels by nitriding and carburizing and in other reactions where small atoms need to penetrate into dense structures. The other mechanisms illustrated, *ring, exchange* and *interstitialcy diffusion*, all require co-operation between several atoms. At first sight, such processes may seem to be less likely than those which involve only

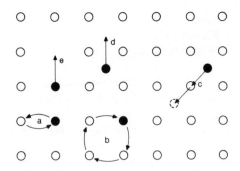

Figure 3.1 Some possible diffusion steps in an idealized crystal structure: (*a*) exchange; (*b*) ring; (*c*) interstitialcy; (*d*) interstitial; (*e*) vacancy.

single particles, but there is considerable evidence to suggest that these mechanisms are by no means rare in practice.

Each time an atom moves, regardless of which mechanism is employed, it will have to overcome a potential energy barrier. This is because the migrating atoms have to leave normally occupied positions which are, by definition, the most stable positions for atoms in the crystal, to pass through less stable positions not normally occupied by atoms. Often atoms may be required to squeeze through a bottleneck of surrounding atoms in order to move at all. Examination of possible diffusion paths in the solid under consideration is needed before the importance of this aspect can be evaluated.

For a one-dimensional diffusion process we can imagine the potential barrier to take the form shown in Figure 3.2. In general, referring to this diagram, we can write H_v for the potential barrier to be surmounted by an atom migrating via a vacancy mechanism, H_i for the potential barrier to be surmounted by an interstitial atom and so on. The symbol a represents the distance between the neighbouring stable sites where the atom can rest. Note that in real diffusion paths H is not necessarily single-valued, and some steps may have a lower value of H than others along the path. Also, it may be preferable to express the energy barrier as a free energy in some circumstances, although we will ignore this extra complexity for the time being.

How easily will an atom overcome the barrier it faces? Obviously, the larger the magnitude of H, the less chance an atom has of making a successful jump. We can gain an estimate of this probability by using classical Maxwell–Boltzmann statistics. Firstly, we need to recall that the atoms in a crystal are not stationary, but are vibrating about the mean position normally occupied. This frequency of vibration, v, is usually taken to have a value of about 10^{13} Hz. Maxwell–Boltzmann statistics then tell us that the probability p that an atom will move from one position of minimum energy (Figure 3.2) to an adjacent position during a vibration in the requisite direction will be given by equation of the following sort:

$$p = \exp(-H/kT) \tag{3.1}$$

Putting some values into equation (3.1), we find, reasonably enough, that if H is very small, the probability approaches 1.0, while if H is equal to kT the

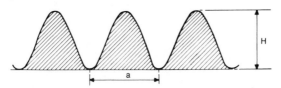

Figure 3.2 Schematic illustration of the potential barrier H that a migrating atom has to overcome in moving through a crystal lattice. Stable positions are separated by the jump distance a.

probability for a successful jump is about one-third*. The number of successful jumps that an atom will make in one second, Γ, will therefore be equal to the attempt frequency, v, multiplied by the probability function above, i.e.:

$$\Gamma = v \exp(-H/kT) \tag{3.2}$$

We will return to this equation later.

3.2.2 Diffusion and atomic migration

As we saw when evaluating experimental diffusion coefficients in Chapter 2, the basic equations of diffusion theory are Fick's laws, which describe the relationships between the flow of diffusing species in a solid and parameters such as time and concentration of diffusing species. For a theoretical analysis of diffusion, the most important of these is *Fick's first law*, which can be written as

$$J = -D\mathrm{d}c/\mathrm{d}x \tag{3.3}$$

where J is the number of particles crossing a unit area in the solid each second and c is the concentration of the diffusing species at point x after time t has elapsed. D is, of course, the diffusion coefficient. Equation (3.3) is for one-dimensional flow, and can be generalized to three dimensions readily. Here though, we will mainly restrict our analysis to one-dimensional diffusion, as this is sufficient to illustrate the theoretical approaches usually made.

We can calculate J theoretically by considering appropriate models for the diffusion process. The simplest model to take corresponds to a situation where we have no significant concentration gradients and the migrating atom takes jumps completely at random. In this case, the atom carries out a *random walk*. This corresponds experimentally to the self-diffusion process. To analyse the process we make use of the situation illustrated in Figure 3.3. Here we have

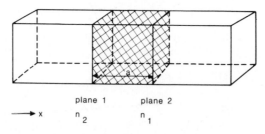

plane 1   plane 2

→ x n_2 n_1

Figure 3.3 Schematic diagram of two adjacent planes 1 and 2 in a crystal, separated by the jump distance for diffusion a. The numbers of diffusing atoms on these planes are n_1 and n_2 per unit area respectively.

*It is well worth taking some time using a calculator to get a feeling for the relationship between p and H.

adjacent lattice planes, numbered 1 and 2, separated by the atomic jump distance, a, shown in Figure 3.2.

Let n_1 and n_2 be the numbers of diffusing atoms per unit area in planes (1) and (2) respectively. If Γ_{12} is the frequency with which an atom moves from plane (1) to plane (2), then the numbers of atoms moving from plane (1) to (2) per unit time is j_{12}, where

$$j_{12} = n_1 \Gamma_{12} \qquad (3.4)$$

Similarly, the number moving from plane (2) to plane (1) is j_{21}, where

$$j_{21} = n_2 \Gamma_{21} \qquad (3.5)$$

The net movement (often called the flux) between the planes, J, is given by

$$J = j_{12} - j_{21} = (n_1 \Gamma_{12} - n_2 \Gamma_{21}) \qquad (3.6)$$

If the process is random, as in self-diffusion, so that the jump frequency is independent of direction we can set Γ_{12} equal to Γ_{21}. Moreover, we have derived an equation for the number of successful jumps that an atom makes per second, viz. equation (3.2). If the jump frequency is independent of direction, then half of the jumps, on average, will be in one direction, and half will be in the opposite direction, so we can write

$$\Gamma_{12} = \Gamma_{21} = \tfrac{1}{2}\Gamma \qquad (3.7)$$

and

$$J = \tfrac{1}{2}(n_1 - n_2)\Gamma \qquad (3.8)$$

To proceed further we must relate n_1 and n_2 to the concentration of mobile atoms in the crystal. This is readily accomplished. Take a single crystal of unit cross-sectional area and unit length. Let the number of mobile atoms in this volume be c. The length of the crystal contains $1/a$ planes, so the concentration of mobile atoms per unit area on each plane will be $c/(1/a)$, that is ca. Let us look now at the situation shown in Figure 3.3, in which we have two neighbouring planes (1) and (2), with populations of mobile atoms n_1 and n_2 and concentrations c_1 and c_2, separated by a distance a. From the discussion above we know that the number of mobile atoms on plane (1) will be n_1, which will be equal to ac_1, and the number of mobile atoms on plane (2) will be n_2, which will be equal to ac_2, and so we can write

$$(n_1 - n_2) = a(c_1 - c_2) \qquad (3.9)$$

Hence

$$J = \tfrac{1}{2}a(c_1 - c_2)\Gamma \qquad (3.10)$$

The concentration gradient, dc/dx, is given by the change in concentration between planes (1) and (2) divided by the distance between planes (1) and (2), which can be expressed algebraically by

$$-dc/dx = (c_1 - c_2)/a \qquad (3.11)$$

where a minus sign is introduced as the concentration falls as we move from plane (1) to plane (2). Hence, from equation (3.11) we can write

$$(c_1 - c_2) = -a\,dc/dx \tag{3.12}$$

and

$$J = -\tfrac{1}{2}\Gamma a^2 dc/dx \tag{3.13}$$

If we now compare equations (3.4) and (3.13) we can write

$$D = \tfrac{1}{2}\Gamma a^2 \tag{3.14}$$

We have already derived an expression for the jump frequency, Γ, in terms of the barrier height to be negotiated, H, so that from equation (3.3) we ultimately find

$$D = \tfrac{1}{2}a^2 v(-H/kT) \tag{3.15}$$

for one-dimensional self-diffusion.

3.2.3 Self-diffusion in crystals

So far our considerations have been limited to a one-dimensional situation. We could generalize all the equations, starting with equation (3.4), for it is clear that in general the diffusion coefficient itself will depend upon direction chosen by the migrating atoms. A treatment of this matter is beyond the scope of this book, but we can take some account of the three-dimensional nature of the diffusion process in the following way. We have so far discussed the jump frequency, Γ, which applied to movement along one direction only, say x. In a three-dimensional structure we can call the overall jump frequency Q. In this case the jump frequency along x will be related to Q by a factor depending on crystal geometry, which we can express as

$$Q = 1/Z \tag{3.16}$$

where Z is the number of jump directions available to a diffusing atom at each step. For example, in a simple cubic system, we have three equivalent directions, so that

$$Q = 1/3 \tag{3.17}$$

and equation (3.15) becomes

$$D = \tfrac{1}{6}a^2 v \exp(-H/kT) \tag{3.18}$$

In general this geometrical constraint can be taken into account by including a factor, g, in equation (3.15), thus:

$$D = ga^2 v \exp(-H/kT) \tag{3.19}$$

The determination of a realistic geometrical factor from *a priori* consideration of crystal structures can be difficult for all but the simplest cases.

There is another modification to equation (3.15) that we can also consider at this stage, and this again derives from a consideration of jump frequency. In a real solid, containing a population of defects, the number of jumps per second will not only involve the diffusing species, but also the defect population. Let us consider two examples. If we are discussing interstitial diffusion in AgBr, the amount of diffusion will be decided by the number of Frenkel defects in the system. If there are no Frenkel defects, there are no interstitial atoms to diffuse at all. In such an event, equations (3.15) or (3.19) must now incorporate a term for n_f. Similarly, an atom at a normal lattice position cannot diffuse by a vacancy mechanism unless there is a vacancy population in the crystal. In a pure crystal, this vacancy population will arise from a Schottky defect population, and we would need to include in equations (3.15) or (3.19) a term for n_s.

The defects of interest could arise in other ways, and, as we shall see in Chapter 4, the population of vacancies and interstitial atoms may be strongly dependent upon impurities present in the materials. If we therefore simply express the number of important defects present in the crystal as n, we can more correctly write equation (3.19) as

$$D = ga^2 vn \exp(-H/kT) \tag{3.20}$$

If we regard g, a^2, v and n as constants these equations are of the same form as equation (2.7), viz.:

$$D = D_0 \exp(-E/RT) \tag{2.7}$$

where D_0, in our analysis is given by $gnva^2$, and the activation energy, E, is equal to the height of the potential barrier, H.

We have therefore succeeded in our aim of relating the observed form of the diffusion equation determined experimentally with the idea of atom jumps constrained by physical parameters of the crystal structure. It is not difficult to see ways in which this approach could be improved, and it is in this way that more elaborate theories of the diffusion process are derived. Some of these refinements will be described later.

3.2.4 The effect of temperature

When the experimentally obtained Arrhenius plots were described in section 2.2, it was remarked that these sometimes fell into two regions; the low-temperature part having a rather lower activation energy and the high-temperature area having a higher activation energy, as is shown schematically in Figure 3.4. Clearly something is happening at higher temperatures which is using additional energy compared to the low-temperature regime. Equation (3.20) gives some indication as to how we can explain this phenomenon. We notice in these equations that the term before the exponential includes n, the number of defects present. At low temperatures the number of intrinsic defects

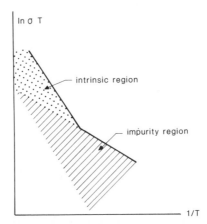

Figure 3.4 Diagram showing the Arrhenius plot expected from a diffusion experiment to find the activation energy for diffusion. The slope in the impurity region yields a value for the enthalpy of movement of the atoms H_m, and the slope in the intrinsic region yields a value for both the enthalpy of formation of the defects H and the enthalpy of movement H_m.

will be small. In fact, as we shall see later, impurities can create defects, and it is reasonable to suppose that the defects due to impurities far outnumber those due to thermal processes, the Frenkel and Schottky defects described in Chapter 1. Hence the number of defects, n, in equation (3.20), is indeed constant, to a first approximation. Thus, in the low-temperature part of an Arrhenius plot the activation energy, E, will correspond to the enthalpy of defect movement, H, as given in equation (3.20) and illustrated schematically in Figure 3.2.

At high temperatures, however, it is clearly unrealistic to assume that the number of defects, n, is constant. Indeed, we have already derived formulae for the variation in the number of Schottky and Frenkel defects with temperature in a crystal of composition MX. It is reasonable, therefore, to assume that at high enough temperatures, the constant n in equation (3.20) should be replaced by a formula expressing the real population of defects in the crystal. Assuming either only Schottky or only Frenkel defects are of importance in this high-temperature regime, we know the number of defects produced at any temperature will be given by equation (1.23) or (1.39). Hence we can expand the formula for D to include an expression for defect numbers, in the following way.

For Frenkel defects we rewrite equation (3.20) as

$$D = gva^2\sqrt{(NN^*)}\exp(-H_i/kT)\exp(-\Delta H_f/2kT) \qquad (3.21)$$

where H_i represents the potential barrier to be surmounted by an interstitial atom. For Schottky defects we obtain the analogous expression

$$D = gva^2 N \exp(-H_v/kT)\exp(-\Delta H_s/2kT) \qquad (3.22)$$

where H_v represents the potential barrier to be surmounted by an atom moving into a vacant site. Both of these equations retain the form of equation (2.7), viz.

$$D = D_0 \exp(-E/RT) \qquad \text{(2.7), (3.23)}$$

but now E is the sum of the energy needed to move the defect, H_i or H_v, plus the energy of defect formation, ΔH_f or ΔH_s, that is, for Frenkel defects

$$E = H_i + \tfrac{1}{2}\Delta H_f \qquad \text{(3.24)}$$

and for Schottky defects

$$E = H_v + \tfrac{1}{2}\Delta H_s \qquad \text{(3.25)}$$

Thus the high-temperature part of the curve will allow one to estimate both of these energy terms.

It is possible, therefore, to use Arrhenius curves which show both intrinsic and extrinsic regions to determine the enthalpy of formation of point defects, and their enthalpy of motion. Some values found in this way are listed in Table 3.1.

Finally, we should remark on the fact that in our previous discussion we have supposed that the height of the potential barrier will be the same at all temperatures. This is obviously not so. As the temperature increases the lattice will expand, and in general H would be expected to decrease. The diffusion coefficient will increase due to this change. Moreover, some of the other constant terms in the preceding equations will vary slightly with temperature. For example, the lattice spacings will change, leading to a change in the constant a, and the vibration frequency, v, will alter somewhat. The Arrhenius plots should reveal this by being slightly curved. We will not take this analysis further here, but note that these effects are generally much smaller than the change between the low-temperature impurity regime and the high-temperature intrinsic region discussed above.

Table 3.1 Some enthalpy values for the formation and movement of vacancies and interstitials in alkali halide crystals

Material	H_f	Schottky defects H_m		Frenkel defects H_m	
		Cation vacancy	Anion vacancy	Interstitial	vacancy
NaCl	192	84	109		
NaBr	163	84	113		
KCl	230	75	172		
KBr	192	29	46		
AgCl	155			13	36
AgBr	117			11	23

All values in kJ mol^{-1}.

3.3 The relationship between D and diffusion distance

In many processes, both industrial and in the laboratory, it is necessary to be able to estimate the distance that an atom can diffuse in a given time. This is important, for example, if we want to know how far a dopant will penetrate into a semiconductor during fabrication. Similarly, we may want to examine the possibility of retaining a high-temperature phase by quenching, and so need to know how far defects can migrate during the period of cooling. To do this we need to have available a relationship between the diffusion coefficient and the distance x that an atom, ion or defect moves in time t, at a given temperature.

Let us consider the movement of an atom from one stable site to the next. The net displacement of a diffusing atom after N jumps will be the algebraic sum of the individual jumps. If x_i is the distance moved along the x axis in the ith jump, the distance moved after a total of N jumps, x, will simply be the sum of all the individual steps, i.e.

$$x = x_1 + x_2 + x_3 \ldots = \sum_{i=1}^{N} x_i \tag{3.26}$$

In our case each individual value of x_i can be $+a$ or $-a$.

If the jumps take place with an equal probability in both directions and the individual jump distances are equal, then this algebraic sum must be equal to zero. This does not mean that the diffusing atom remains at its starting point after N jumps but only that the jumps in a positive and negative direction are equally probable. In fact the total displacement may have any value between zero and Na.

In order to proceed we use a mathematical trick. If the jump distances are squared, we automatically get rid of all the negative quantities. So to square equation (3.26) we write

$$x^2 = (x_1 + x_2 + x_3 \ldots x_N)(x_1 + x_2 + x_3 \ldots x_N)$$

$$= (x_1 x_1 + x_1 x_2 + x_1 x_3 \ldots$$

$$+ x_2 x_1 + x_2 x_2 + x_2 x_3 \ldots$$

$$+ \ldots$$

$$+ x_N x_1 + x_N x_2 + \ldots) \tag{3.27}$$

We can write this in a more condensed form as

$$x^2 = \sum_{i=1}^{N} x_i^2 + 2 \sum_{i=1}^{N-1} x_i x_{i+1} + 2 \sum_{i=1}^{N-2} x_i x_{i+2} + \cdots$$

$$= \sum_{i=1}^{N} x_i^2 + 2 \sum_{j=1}^{N-1} \sum_{i=1}^{N-1} x_i x_{i+j} \tag{3.28}$$

This does not appear to be any simpler than our starting equation, but if we consider a large number of jumps, and knowing that each jump may be either positive or negative, the double sum term in equation (3.28) averages to zero. This is not intuitively obvious, but some small investment of time to carry out the calculation with a variety of values for N will show that this will be true, as $x_i x_{i+j}$ is just as likely to be positive as negative. This procedure is equivalent to deriving the average of the square of the displacement over many jumps, and the result is called the *mean square displacement*, and written $\langle x^2 \rangle$. The formidable equation (3.27) therefore reduces to the manageable form

$$\langle x^2 \rangle = \sum x_i^2 \qquad (3.29)$$

As each jump, x_i, can be equal to $+a$ or $-a$,

$$\langle x^2 \rangle = x_1^2 + x_2^2 + x_3^2 \ldots + x_N^2$$
$$= a^2 + a^2 + a^2 \ldots + a_N^2$$

i.e.

$$\langle x^2 \rangle = Na^2 \qquad (3.30)$$

In section 3.2 we defined Γ as the frequency with which an atom jumps from one site to another along the x direction, so that the total number of jumps, N, will be given by Γ jumps per second multiplied by the time, t, over which the diffusion experiment has lasted, that is

$$N = \Gamma t \qquad (3.31)$$

Hence

$$\langle x^2 \rangle = \Gamma t a^2 \qquad (3.32)$$

However, equation (3.14) shows that the term Γa^2 can be replaced by twice the self-diffusion coefficient, D, so that

$$\langle x^2 \rangle = 2Dt \qquad (3.33)$$

The average distance that an atom will travel in time t is then the square root of $\langle x^2 \rangle$, a quantity called the *root mean square* value of x, which is given by

$$\sqrt{\langle x^2 \rangle} = \sqrt{2Dt} \qquad (3.34)$$

Thus, we find that the root mean square distance that an atom will move is given by a rather simple relationship, and that it is proportional to the square root of the time.

The derivation just carried out was straightforward, and designed to show the general approach to the problem of relating the distance that a diffusing atom will move in a certain time to the diffusion coefficient. As each jump is not connected to the previous one or the following one the process is called a *random walk* analysis. However, the simplicity of the approach disguises the fact that we must be rather careful about exactly what equation (3.34) means. We wanted to calculate how far an atom would diffuse in a time t, and have ended up with an average distance called the root mean square distance. Is this

c

the solution we need? Let us consider the situation in a little more depth.

Some thought will show that there is no single answer to the problem that we set out to solve in this section. Recall the experimental determination of self-diffusion coefficients in the previous chapter. In this, the first task was to measure the penetration curve for the diffusion of a tracer atom in the host crystal. What, precisely, does this curve tell us? The curve relates the concentration of diffusing atoms to the distance from the original interface in a diffusion couple. That is, the graph already shows us that some diffusing atoms have travelled a fairly long way during the experiment, while others have not gone far, exactly as the summation in equation (3.26) revealed. There is no single answer, therefore, to the question of how far an atom can diffuse in a given period of time. The best that we can do is to obtain some measure of the average distance, or more properly, the most likely distance, that an atom will travel during the time of the diffusion experiment. This is what equation (3.34) provides, but we are still no nearer understanding the significance of the root mean square displacement. To answer this it is necessary to turn to the statistics of atom movements.

If we have motion consisting of random atom jumps along the x-axis, as we do in self-diffusion, a statistical analysis* of the situation will show that the concentration profile for the diffusing atoms will be a bell-shaped curve of the type shown in Figure 3.5. Figure 2.2 shows such bell-shaped curves, in fact, although only half of the curves were drawn in the diagram. In statistical parlance the distribution of atoms leading to a bell-shaped curve is called a *normal distribution* or *Gaussian distribution*. The statistics of the normal distribution are well known and we can use them to throw some light on the exact meaning of the root mean square displacement. It is found that the probability P that after N jumps an atom will be displaced a distance x is given, for our one-dimensional case, by the equation

$$P(x, N) = [\tfrac{1}{2}\pi N a^2]^{1/2} \exp[-x^2/2Na^2] \qquad (3.35)$$

where a is the jump distance, defined earlier. The form of this function closely resembles the solution to the tracer diffusion equation given in the previous chapter.

Now we know that if the jump frequency, Γ, is the same for diffusion to the left or right, the total number of jumps in either direction, N, is given by

$$N = \Gamma t \qquad (3.36)$$

---

*It is interesting to learn that the equations relevant to this analysis were first derived by Demoivre in 1733, with respect to problems associated with tossing of coins. There are many similarities, of course. Each atom can jump forwards or backwards, just as a coin can fall 'heads' or 'tails'. A very reasonable model for diffusion can be made simply by placing a row of counters on a board, and tossing a coin to decide if a counter should move forward or backward. After many throws the distribution of the counters will mirror the distribution of atoms. Of course such a procedure is now most easily carried out using a computer routine.

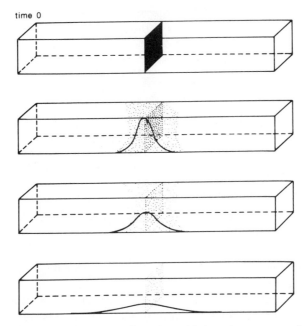

time 0

Figure 3.5 Development of bell-shaped curves by diffusion of a tracer atom in a crystal.

where t is the time of reaction. Also remembering that

$$D = \tfrac{1}{2}\Gamma a^2 \qquad\qquad (3.14), (3.37)$$

we can write equation (3.35) as

$$P(x, t) = [\tfrac{1}{2}\sqrt{(\pi Dt)}] \exp[- x^2/4Dt] \qquad\qquad (3.38)$$

where $P(x, t)$ is the probability that after time t the atom will be found at position x. This is now, except for the concentration factor, identical to equation (2.3). This is easily added because, if we start with all our atoms on a plane at $x = 0$ at time $t = 0$ with a concentration of c_0 atoms per unit area, then the concentration after time t at a position x given by $c(x, t)$, where

$$c(x, t) = c_0 P(x, t) \qquad\qquad (3.39)$$

that is

$$c(x, t) = [c_0/2\sqrt{(\pi Dt)}] \exp[- x^2/4Dt] \qquad\qquad (3.40)$$

thus reproducing equation (2.3) exactly.

A plot of equation (3.40) produces a bell-shaped normal distribution curve, and the form of the equation means that each point on the curve represents the probability, P, of finding a diffusing atom in the small interval $x + \mathrm{d}x$. The probability that an atom will be found in a given region of crystal between 0 and $\pm x$ will then be equal to the area under the curve between the limits chosen, viz. $\pm x$.

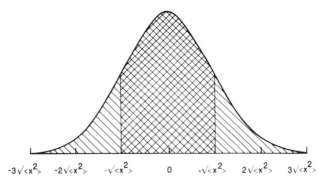

Figure 3.6 A typical bell-shaped normal distribution curve. The shaded region in the centre of the curve shows the area between the limits $+\sqrt{\langle x^2 \rangle}$ and $-\sqrt{\langle x^2 \rangle}$ which must be evaluated to determine the probability that a diffusing atom will be found in the same region of material. This is found to be equal to 67%.

We can use this to finally determine the strict meaning of the root mean square displacement. In Figure 3.6 we have chosen the value $\sqrt{\langle x^2 \rangle}$ as our value of x. The hatched area thus gives the probability that an atom will be found in the volume of solid bounded by $\pm\sqrt{\langle x^2 \rangle}$. This is found to be 0.67. Thus, the real meaning of $\sqrt{\langle x^2 \rangle}$ is that there is a 67% probability that any particular atom will be found in the region between the starting point of the diffusion and a distance of $\pm\sqrt{\langle x^2 \rangle}$ on either side of it. The probability that any particular atom has diffused *further* than this distance is given by the total area under the curve, always normalized to 1.0, minus the shaded area. This is 0.33, that is, 33%. Moreover, the probability that the atoms have diffused further than $2\sqrt{\langle x^2 \rangle}$ is equal to the total area under the curve, 1.0, minus the area under the curve up to $2\sqrt{\langle x^2 \rangle}$. This is found to be equal to about 0.05, that is, about 5%. Some atoms will have gone further than this distance, but the probability that any one particular atom will have done so is very small.

The equations considered here have concentrated on one-dimensional diffusion examples, so as to emphasize the principles involved in the analysis. It does not require a great deal of effort to expand them to cover more complex cases, and examples will be found in the Supplementary Reading listed in section 3.7.

3.4 Correlation effects

In this chapter so far we have discussed the diffusion of atoms or defects in a random fashion, that is, self-diffusion. Each step was unrelated to the one before and not driven by any particular force. The atoms and defects can be considered to be shuffling around, jostled solely by thermal energy. However, diffusion of an atom in a solid is not a truly random process and it is reasonable to suppose that in many circumstances a given jump direction may depend on the direction of the previous jump. When this is so, the random walk diffusion equations need to be modified by the introduction of a *correlation factor, f,* which may be regarded as the ratio of the self-diffusion coefficient actually observed for an atom to that which would be expected on the basis of a truly random migration.

We can explain the situation by considering the vacancy diffusion of an atom in a crystal, as shown schematically in Figure 3.7. The atom we are interested in is stippled in Figure 3.7(a), and can be regarded as a tracer atom. It is situated next to a vacant site, so that diffusion can take place. The diffusion can, in fact, occur by way of any of atoms around the vacancy moving into the empty site. The vacancy of course, has no preference for any of its neighbours so that its first jump is entirely random. Because we are interested in the diffusion of the tracer we can reasonably assume that it is the tracer that makes the first jump into the vacant site. This leads to the situation shown in Figure 3.7(b).

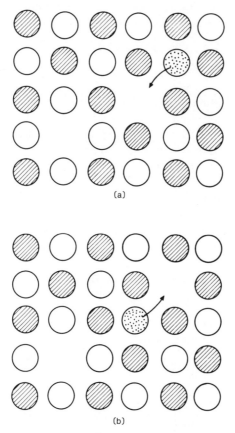

(a)

(b)

Figure 3.7 Correlated motion in vacancy diffusion. The dot shaded circle represents the tracer atom and the curved arrows the most likely next jump of the tracer.

The next jump of the tracer is not, now, an entirely random process. It is still next to the vacancy but the other atoms, also originally neighbours of the vacancy, are no longer in this special situation. Clearly it is more likely that the tracer will move back to the vacancy, recreating the situation shown in Figure 3.7(a) rather than exchange places with a cation nearby, for in this latter case the vacancy must move to the desired location before the tracer can make the necessary movement. Hence, of the choices available to the tracer in Figure 3.7(b), a jump back to the situation shown in Figure 3.7(a) is of highest probability. Note, however, that this argument does not apply to the vacancy, which can always move to an adjacent cation site, and hence can follow a truly random path.

When this non-random motion is considered over many jumps, the mean square displacement of the tracer will be less than that of the vacancy, which took the same number of jumps. This means that the observed diffusion coefficient of the tracer will be less than that of the vacancy. The correlation factor is given by the ratio of the values of the mean square displacement, provided that the number of jumps considered is large. Hence

$$f = \langle x^2 \rangle_{\text{tracer}} / \langle x^2 \rangle_{\text{vacancy}} \tag{3.41}$$

Correlation factors for vacancy diffusion generally take values of between 0.5 and 0.8.

Of course, correlation factors are not limited to vacancy diffusion. If we consider interstitial diffusion in which we have only a few diffusing ions and many available empty sites, we would expect a correlation factor close to 1.0, but for interstitialcy diffusion, this will not be true. A number of mathematical procedures have been adopted for evaluating correlation factors. Table 3.2 lists some values for a variety of diffusion mechanisms in some common crystal structure types.

Before concluding this section, it is useful to take the arguments a little further. In our discussion of vacancy diffusion all the cations were assumed to be identical, and we have a situation similar to that described in section 2.2.1,

Table 3.2 Correlation factors for self-diffusion

Mechanism	Structure	Correlation factor (f)
Vacancy	Diamond	0.50
	b.c.c.	0.7272
	f.c.c.	0.7815
	h.c.p.	$0.7812(f_x, f_y)$
	h.c.p.	$0.7815(f_z)$
Interstitialcy	f.c.c	0.80
	Fluorite (cation diffusion)	1.00
	b.c.c.	0.666
	CsCl (cation diffusion)	0.832
	AgBr (cation diffusion)	0.666

where Mg diffuses through MgO. Often, however, we need to consider the diffusion of an impurity atom in a crystal, say K in NaCl or Ca in MgO. In such cases the probability that the impurity will exchange with the vacancy will depend on other factors such as atomic size or, in the case of ionic movement, the charge on the diffusing species. This can be expressed in terms of the jump frequencies of the host and impurity atoms, in which case one is likely to be greater than the other. If, for example, the host atoms have a very high jump frequency, they will be far more likely to move at any instant, giving them a higher correlation factor than the impurities.

3.5 Diffusion in a potential gradient

3.5.1 *Ionic conductivity*

The potential barrier that an atom must surmount during diffusion can be easily modified by external factors. Of these, electric fields must be considered of prime importance in view of the fact that they influence the diffusion of ions enormously. Ionic movement under the action of an externally applied electric field is called *ionic conductivity*. Although at first sight it may seem that ionic conductivity has little connection with self-diffusion, we will see that the processes can be treated in very similar ways.

Let us, to illustrate this point, consider diffusion of monovalent ions. The potential energy barrier H, to be surmounted by the ions in migrating, will be modified in the presence of an applied field E in the way shown schematically in Figure 3.8. In the situation considered here, the change in the height of the potential barrier will be eaE, where e is the charge on the monovalent ion, a is the separation of the stable resting positions of the ions and E is the magnitude of the applied electric field. However, the most important aspect of the new situation is that the potential barriers are now tilted in the direction of the applied electric field as shown in Figure 3.8. As can be seen, in the direction of

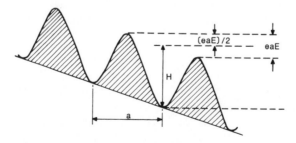

Figure 3.8 Schematic illustration of the potential barrier that a migrating ion must overcome in the presence of an electric field. The values of a, the jump distance, and H, the average height of the potential barrier, are the same as in Figure 3.2, but the effective barrier that an ion faces is lowered for movement in the direction of the field and increased for movement in a direction against the field.

the applied field the barrier is reduced by an amount $eaE/2$ while in a direction against the field it is increased by a similar amount. At right angles to the field, H remains unchanged. Migration of charged particles in the direction of the field is now obviously favoured, as the potential barrier is lower, whilst migration in the opposite direction is less likely.

We can consider the effect this will have upon the movement of the ions by utilizing the same model as we employed in section 3.2.2. Hence, the number of jumps that an ion will make in the direction of the field per second is given by a modified form of equation (3.2), viz.

$$\Gamma_+ = v \exp\left[-(H - \tfrac{1}{2}eaE)/kT\right] \tag{3.42}$$

where we have simply substituted the new potential barrier $H - \tfrac{1}{2}eaE$ for H. In a direction against the field the number of jumps will be given by the analogous expression

$$\Gamma_- = v \exp\left[-(H + \tfrac{1}{2}eaE)/kT\right] \tag{3.43}$$

where we have a similar substitution for H but this time of $H + \tfrac{1}{2}eaE$, as the barrier height is now increased.

The overall jump rate in the direction of the field is $\Gamma_+ - \Gamma_-$, and as the net velocity of the ions in the direction of the field, V, is given by the net jump rate multiplied by the distance moved at each jump, we can write

$$V = va \exp\left[-(H - \tfrac{1}{2}eaE)/kT\right] - va \exp\left[-(H + \tfrac{1}{2}eaE)/kT\right] \tag{3.44}$$

$$= va \exp(-H/kT)[\exp(eaE/2kT) - \exp(-eaE/2kT)] \tag{3.45}$$

For low field strengths, eaE is much less than kT, and $\exp(eaE/2kT) - \exp(-eaE/2kT)$ may be replaced by eaE/kT, as we see from Table 3.3. To see what sort of field strength this approximation corresponds to, we can take a value of $eaE/2kT$ equal to 1 to be the maximum value at which the approximation holds and estimate the corresponding field strength. Thus, taking a temperature of 500 K, and a value of a of about 0.3×10^{-9} m, substitution into the equation yields a value for E of 2.87×10^2 V cm^{-1}. Thus, the approximation is a reasonable one for the field strengths up to about 300 V cm^{-1}.

Table 3.3 The equivalence of eaE/kT and $[\exp(eaE/2kT) - \exp(-eaE/2kT)]$

$eaE/2kT$	$\exp(eaE/2kT)$	$\exp(-eaE/2kT)$	Difference
0.001	1.0010	0.999	.002
0.01	1.0101	0.999	.020
0.10	1.1052	0.9048	.2003
1.0	2.7183	0.3679	2.3504
10.0	22026.47	4.54×10^{-5}	22026.47

We can now proceed to write

$$V = (va^2 eE/kT)\exp(-H/kT) \qquad (3.46)$$

The mobility μ of the ion is the rate of movement when the value of E is unity, i.e.

$$\mu = (va^2 e/kT)\exp(-H/kT) \qquad (3.47)$$

Finally we arrive at an expression for the ionic conductivity, which is given by

$$\sigma = ne\mu \qquad (3.48)$$

where n is the number of migrating ions per unit volume. We can therefore write

$$\sigma = (nva^2 e^2/kT)\exp(-H/kT) \qquad (3.49)$$

We can see that this equation takes on a form

$$\sigma = (\sigma_0/T)\exp(-H/kT) \qquad (3.50)$$

which is very similar to the diffusion equation that was derived earlier. Indeed, we would expect this to be so, as we have used the same method of mathematical analysis as before. The only difference that the field makes is in the term equivalent to D_0 which is now replaced by σ_0/T.

In equation (3.49) the pre-exponential factor includes a term n which corresponds to the number of mobile species present in the crystal. In some cases, such as when atom migration takes place via a vacancy diffusion mechanism, n will be closely related to the population of point defects present. At low temperatures, where this population is controlled by the impurities present, n will be constant, and we can write, for vacancies,

$$\sigma_v = [n_v va^2 e/kT]\exp(-H_v/kT) \qquad (3.51)$$

and for interstitials

$$\sigma_i = [n_i va^2 e/kT]\exp(-H_i/kT) \qquad (3.52)$$

In this regime we can obtain measures of H_v or H_i directly from the Arrhenius plots of $\ln \sigma T$ verus $1/T$. At high temperatures it is reasonable to suppose, as in the case of diffusion, that the values of n_v or n_i are temperature-dependent. In this case we can substitute for n from the equations given in Chapter 1 to obtain, for interstitials due to Frenkel defects

$$\sigma_i = [va^2 e^2 (NN^*)^{1/2}/kT]\exp(-H_i/kT)\exp(-\Delta H_f/2kT) \qquad (3.53)$$

and for vacancies due to Schottky defects

$$\sigma_v = [va^2 e^2 N/kT]\exp(-H_v/kT)\exp(-\Delta H_s/2kT) \qquad (3.54)$$

In this case, an Arrhenius plot of $\ln \sigma T$ versus $1/T$ will yield a higher value for the activation energy than in the low-temperature region. The new value for E

will be composed of two terms, viz.

$$E_s = H_v + \tfrac{1}{2}\Delta H_s \qquad (3.55)$$

for Schottky defects, and

$$E_f = H_i + \tfrac{1}{2}\Delta H_i \qquad (3.56)$$

for Frenkel defects. If both high- and low-temperature regimes are important, the Arrhenius plots for such materials will show a knee similar to that in Figure 3.4.

Thus one is able to measure values for enthalpy of formation and migration of point defects using conductivity data in an analogous way to diffusion data. In fact, although conductivity measurements are not always easy to make, they are usually much easier to perform than diffusion experiments, and are preferred if they can give needed information. To illustrate this, Figure 3.9 reproduces some experimental results for the ionic conductivity of NaCl which show the intrinsic and impurity regions clearly.

The equations that have been derived in this section are correct for monovalent ions only. In order to make them applicable to ions of formal charge $+z$, we must replace the term e, representing the charge on the ions, by $+ze$, so that terms such as eaE become $zeaE$ in all that follows. Hence

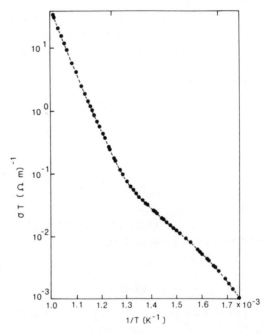

Figure 3.9 The ionic conductivity of NaCl, clearly showing the intrinsic and impurity regions. From R. Kirk and P.L. Pratt, *Proc. Brit. Ceram. Soc.* **9** (1967) 215.

equation (3.49), which is the most important equation in the section, becomes

$$\sigma = [nva^2z^2e^2/kT]\exp(-H/kT) \qquad (3.57)$$

It would also be possible to include geometrical factors in the equation above, and consider temperature changes in more detail. However, these modifications can be made with reference to the earlier material if they are needed.

3.5.2 The relationship between ionic conductivity and diffusion coefficient

From our foregoing discussions, ionic conductivity and ionic diffusion are seen to be closely related. If both processes occur by the same random walk mechanism, the relationship between the self-diffusion coefficient, D, and the ionic conductivity, σ, can readily be derived.

In the previous section the ionic conductivity of a monovalent ion in a direction parallel to an external electric field was given by equation (3.49), i.e.

$$\sigma = [nva^2e^2/kT]\exp(-H/kT) \qquad (3.49)$$

The equivalent equation for diffusion of an ion, moving in one direction, over an identical potential barrier will be given by

$$D = va^2\exp(-H/kT) \qquad (3.58)$$

Combining these two equation gives

$$\sigma/D = ne^2/kT \qquad (3.59)$$

This equation, first derived by Albert Einstein, is a simplified form of an equation generally known as the Nernst–Einstein equation. As in the case of the conductivity equations given above, for an ion of valence $+z$ the equation becomes

$$\sigma/D = nz^2e^2/kT \qquad (3.60)$$

In both these equations n is the number of mobile ions of charge ze per unit volume.

It is revealing to estimate the value of σ/D. If σ is measured in ohm^{-1} cm^{-1},

$$\sigma/D = n \times (1.6 \times 10^{19})^2/(1.38 \times 10^{23})T$$
$$= 1.86 \times 10^{15}n/T$$

In general n takes a value of approximately 10^{22} defects cm^{-3}, and, taking T as 600 K,

$$\sigma/D = 3.1 \times 10^4 \qquad (3.61)$$

Hence we see that D is considerably smaller than σ, so not only are conductivity values more easily obtained experimentally, but they are also a lot larger in magnitude.

Although some care must be exercised in using the equations given in this section, as they assume that exactly the same mechanism holds for both the ionic conductivity and the diffusion of the charged species, they are extremely useful, as they can be used to determine self-diffusion coefficients from the more easily obtained conductivity data. However, when experimental values of both conductivity and diffusion coefficient are available, if it is found that values of the diffusion coefficient calculated from conductivity differ from that measured directly, it suggests that the mechanisms of the two processes are different. In some circumstances this difference can yield information about the atomic mechanism of the diffusion process.

3.5.3 Transport numbers

The conventional method of stating the electrical conductivity of a crystal gives no indication of the component conductivities which may be contributing to the overall effect. Conductivity could arise from cations, anions or electrons. *Transport numbers* give the extent to which each of these factors contribute to the conductivity. Thus if we denote the total conductivity of a material by σ we can write

$$\left.\begin{aligned}
\sigma &= \sigma_{\text{cation}} + \sigma_{\text{anion}} + \sigma_{\text{electron}} \\
\sigma_{\text{cation}} &= t_{\text{cation}}\sigma \\
\sigma_{\text{anion}} &= t_{\text{anion}}\sigma \\
\sigma_{\text{electron}} &= t_{\text{electron}}\sigma
\end{aligned}\right\} \tag{3.62}$$

where σ_{cation}, σ_{anion} and σ_{electron} are the conductivities of the cations, anions and electrons and t_{cation}, t_{anion} and t_{electron} are called the transport numbers for cations, anions and electrons respectively. As can be seen from the relationships in equation (3.62),

$$\sigma = \sigma(t_{\text{cation}} + t_{\text{anion}} + t_{\text{electron}})$$

and

$$t_{\text{cation}} + t_{\text{anion}} + t_{\text{electron}} = 1 \tag{3.63}$$

The experimental determination of transport numbers is not always easy, particularly as polarization effects at electrodes can lead to difficulties in the interpretation of the results, and for this reason values given in the literature should be regarded with caution. There are, however, some generalizations that can be made. For example:

(i) Halides rarely show electronic conductivity, and many, including the Li and Ag halides, have $t_{\text{cation}} = 1.0$, while others, such as the Ba and Pb halides, have $t_{\text{anion}} = 1.0$. NaF, NaCl, NaBr and KCl have both cation and anion contributions.

(ii) Oxides and sulphides, especially of transition metals, often have appreciable values of t_{electron}.

3.5.4 *Diffusion in a concentration gradient*

The models that we have used so far to discuss diffusion and ionic conductivity can be generalized to include the case when the potential gradient is due to a change in chemical potential rather than electrical potential. It is of interest to run through the reasoning as it applies to jump frequencies, so as to see how this analysis could be used if need be. Once again we will limit ourselves to one-dimensional matter flow. In the most general case the presence of a gradient in the chemical potential will cause a flow of atoms in a preferred direction. We can then use Figure 3.8 to illustrate this, although now it is necessary to relate the change in the height of the barrier to diffusion to the Gibbs free energy change or to the change in chemical potential. This is shown schematically in Figure 3.10.

The analyses carried out in sections 3.2 and 3.5.1 are now repeated, but with the following amendments. There will be a greater number of successful jumps from left to right than from right to left due to the asymmetrical nature of the barrier that the migrating atoms have to surmount. Suppose that the height of the barrier in the absence of the perturbing chemical potential is G. If a force F results in a change in the potential barrier of the type shown in Figure 3.10, the barrier will be lowered in a direction from left to right by ΔG, where

$$\Delta G = \tfrac{1}{2}aF \tag{3.64}$$

Hence the jump frequency from left to right will be

$$\Gamma_{lr} = \exp[-(G - \Delta G/kT) \tag{3.65}$$

$$= \tfrac{1}{2}\Gamma_0 \exp(G/kT) \tag{3.66}$$

where Γ_0 is the jump frequency in the absence of the force and the factor of $1/2$ has been introduced to account for the fact that the atoms may jump in either of two directions when Γ_0 is evaluated.

Considering the atom flow from right to left, the potential barrier is now

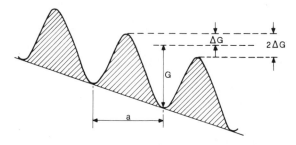

Figure 3.10 Schematic illustration of the barrier that a migrating ion or atom must overcome in a chemical potential gradient. The distance between stable sites is represented by a, and the average barrier by the free energy, G. The barrier that a mobile atom or ion has to overcome is decreased by ΔG in one direction and increased by ΔG in the opposite direction.

increased to $G + \Delta G$. Hence the jump frequency from right to left will be

$$\Gamma_{rl} = \exp\left[-(G + \Delta G)/kT\right] \tag{3.67}$$

$$= \tfrac{1}{2}\Gamma_0 \exp(-\Delta G)/kT \tag{3.68}$$

The net jump rate, Γ, is given by

$$\Gamma = \Gamma_{rl} - \Gamma_{lr} \tag{3.69}$$

$$= \tfrac{1}{2}\Gamma_0 [\exp(+\Delta G)/kT - \exp(-\Delta G/kT)] \tag{3.70}$$

This expression can be simplified provided that ΔG is much smaller than kT. In this case

$$\exp(\Delta G)/kT \approx 1 + (\Delta G/kT) \tag{3.71}$$

$$\exp(-\Delta G)/kT \approx 1 - (\Delta G/kT) \tag{3.72}$$

Hence

$$\Gamma = \tfrac{1}{2}\Gamma_0 (2\Delta G/kT) \tag{3.73}$$

If we substitute for ΔG from equation (3.64), we can write

$$\Gamma = \tfrac{1}{2}\Gamma_0 (Fa/kT) \tag{3.74}$$

The net velocity of the atoms will be given by the net jump rate multiplied by the jump distance, that is

$$V = \tfrac{1}{2}\Gamma_0 (Fa^2/kT) \tag{3.75}$$

The flux of atoms, J, will now be given by the product of the velocity of the atoms multiplied by the number of diffusing atoms per unit volume, n, that is

$$J = \tfrac{1}{2}\Gamma_0 (Fa^2 n/kT) \tag{3.76}$$

If we finally substitute for Γ_0,

$$\Gamma_0 = \exp(-G/kT) \tag{3.77}$$

$$J = \tfrac{1}{2}(Fa^2 n/kT)\exp(-G/kT) \tag{3.78}$$

It is seen that this equation is identical in form to that of equation (3.46) if the force F is equated with the electrical force eE. We can, more generally, consider that the force is due to a gradient in electrochemical potential and derive related expressions for such. In this case both ionic conductivity and chemical diffusion will be linked.

3.6 Conclusions

At the outset of this chapter we tried to use simple ideas to see if we could gain an insight into the relationship between the diffusion coefficient and the atomic mechanisms by which atoms moved through crystals. In this we have been remarkably successful, because we have also gained an insight into the process of ionic conductivity, which one would not intuitively link with diffusion. In addition, it is clear that our analysis could readily be extended to more complex situations by removing some of the restrictions that we have imposed for the sake of simplicity.

3.7 Supplementary reading

Atomic diffusion is covered in many textbooks of materials science and physical chemistry. For a very clear, but introductory, article, see *Phase Diagrams and Microstructure*, Open University Introduction to Materials course, Unit 5, Open University (1974).

A more comprehensive discussion is included in W.D. Kingery, H.K. Bowen and D.R. Uhlmann, *Introduction to Ceramics*, 2nd edn., Wiley-Interscience (1976).

At an advanced, but still readable level, refer to P.G. Shewman, *Diffusion in Solids*, McGraw-Hill (1963); J.R. Manning, *Diffusion Kinetics for Atoms in Crystals*, Van Nostrand (1968).

A clear exposition of the principles of diffusion, together with self-assessment questions, is given by R. Metselaar, *J. Mater. Ed.* **6** (1984) 229; **7** (1985) 653.

The Supplementary Reading at the end of Chapter 2 also contains material relevant to the present chapter.

Part 2 Non-stoichiometric materials containing ions with fixed valence

4 Non-stoichiometry and defect chemistry

4.1 The composition of solid phases

Classical chemistry has taught people to think in terms of compounds in which the ratios of the atomic components are small integers. The compositions of such compounds are, moreover, thought to be totally invariant. Examples are given by molecules such as HCl, H_2O and NH_3, or by solids which are made up of molecular components, typified by organic crystals. Experimentally it is found that these sorts of material have exact and fixed compositions. They are known as *stoichiometric compounds*.

This concept, which is embodied in the quantitative laws of chemistry, does not apply very well to numerous solid phases. A careful examination of the compositions of solids, particularly inorganic compounds, frequently shows that the ratios of the atomic components are not integers. Such compounds occur in many different families of inorganic materials. A typical example is given by vanadium monoxide, which can, in fact, take any composition between the approximate limits of $VO_{0.8}$ to $VO_{1.27}$, and does not just exist at the precise composition $VO_{1.0}$. Such compounds are termed *non-stoichiometric*, and the example quoted reveals two characteristics of such compounds; firstly, they invariably have irrational formulae, and secondly, they are found to possess appreciable composition ranges. In order to make it clear when we are talking of non-stoichiometric phases, their formulae will be prefaced by the symbol $\approx$.

In order to understand why non-stoichiometric phases such as $\approx VO$ should form we need to turn to the equivalent question, why do 'normal' compounds have a fixed composition with no latitude at all? The answer to this lies in the rules of chemical valence, and the valence requirements of the atoms within the compound in question. Thus, we need to recall the elements of bonding theory to go further.

Let us consider molecules again. Molecules are usually held together by strong covalent bonds which are formed by the pairing of two electrons. A great deal of energy is required to break these bonds, and the consequence is that in normal circumstances molecules are stable and show fixed compositions which are simply related to the number of available unpaired valence electrons. The same would therefore be expected to be true for solids which are made up of a packing of molecules; these would be expected to be stoichiometric and not to show a composition range in normal circumstances. Diamond, and other crystals which contain atoms linked by a strong network of covalent bonds, would similarly be expected to show no composition range.

The bonding electrons in ionic crystals are also localized. The removal or addition of ions would require the expenditure of a great deal of energy and also leave the crystals in a charged state. Hence ionic crystals would also be expected to be stoichiometric phases. There is one possible exception to this statement that can be envisaged, though. If the crystal contains ions which can take on several valences, then it is possible that a change in the number of ions present could be compensated for by a variation in the valence of the ions in the crystal which would overcome the problems of charge balance. Thus we would consider that, say, transition metal compounds would be more likely to show a composition range than compounds of non-transition metals.

When we turn to metals the situation is quite the opposite. In a metal the outer bonding electrons are spread throughout the crystal. Any element would therefore be expected to be able to enter the structure easily in any amount, provided that it could add its electrons to the communal pool. Alloys with extended phase ranges are therefore the rule rather than the exception. Alloys, of course, do have structures and specific composition ranges, but these are often controlled by band-filling considerations, which lead to the idea of electron compounds and the Hume–Rothery rules which explain the composition ranges of such phases. In general, though, it is not so surprising, in bonding terms, that compounds using metallic bonding are non-stoichiometric in behaviour.

These statements are, of course, gross simplifications, albeit useful ones, and so must be treated cautiously. For example, a factor that must always be taken into account is temperature. As the temperature increases, the bonding liaisons between atoms in a solid weaken, and we progress to liquid or vapour states. Thus the generalizations above would be expected to weaken at higher temperatures. This is particularly so when the liquid state is reached. As glasses are supercooled liquids, we would thence expect glassy solids to show composition ranges, and this is frequently found to be the case.

These introductory considerations indicate that the non-stoichiometric phases most commonly met are likely to be metallic alloys and transition metal compounds, especially oxides, sulphides, carbides, nitrides and hydrides. In this volume we intend to focus attention upon inorganic non-metallic materials, and so a short but by no means exhaustive list of some of these non-stoichiometric phases is given in Table 4.1.

Before closing this section it is useful to note that many non-stoichiometric compounds possess useful technological properties. Moreover, these can be modified by varying the relative proportions of the atomic constituents. This is understandable from the bonding considerations outlined above. A change in the number of atoms present in the solid will alter the number of incomplete

Table 4.1 Approximate composition ranges for some non-stoichiometric phases*

Compound		Composition range
TiO_x	[≈ TiO]	$0.65 < x < 1.25$
	[≈ TiO_2]	$1.998 < x < 2.000$
VO_x	[≈ VO]	$0.79 < x < 1.29$
Mn_xO	[≈ MnO]	$0.848 < x < 1.00$
Fe_xO	[≈ FeO]	$0.833 < x < 0.957$
Co_xO	[≈ CoO]	$0.988 < x < 1.000$
Ni_xO	[≈ NiO]	$0.999 < x < 1.000$
CeO_x	[≈ Ce_2O_3]	$1.50 < x < 1.52$
	[≈ $Ce_{32}O_{58}$]	$1.805 < x < 1.812$
ZrO_x	[≈ ZrO_2]	$1.700 < x < 2.004$
UO_x	[≈ UO_2]	$1.65 < x < 2.25$
$Li_xV_2O_5$		$0.2 < x < 0.33$
$Na_xV_2O_5$		$0.13 < x < 0.31$
$Li_xV_3O_8$		$1.13 < x < 1.33$
Li_xWO_3		$0 < x < 0.50$
Ca_xWO_3		$0 < x < 0.125$
In_xWO_3		$0.20 < x < 0.33$
[Pb, Bi]F_x	[$PbF_2 - Bi_2F_3$]	$2.0 < x < 3.0$
TiS_x	[≈ TiS]	$0.971 < x < 1.064$
	[≈ Ti_8S_9]	$1.112 < x < 1.205$
	[≈ Ti_3S_4]	$1.282 < x < 1.300$
	[≈ Ti_2S_3]	$1.370 < x < 1.587$
	[≈ TiS_2]	$1.818 < x < 1.923$
Nb_xS	[≈ NbS]	$0.92 < x < 1.00$
Ta_xS_2	[≈ TaS_2]	$1.00 < x < 1.35$
$Ba_3Fe_{1+x}S_5$		$0 < x < 1$
Y_xSe	[≈ YSe]	$1.00 < x < 1.33$
Ni_xSe	[≈ NiSe]	$0.77 < x < 0.82$
V_xTe_2	[≈ VTe_2]	$1.03 < x < 1.14$
Mo_xTe_2	[≈ $MoTe_2$]	$1.00 < x < 1.05$

*Note that all composition ranges are temperature-dependent and the figures given here are intended only as a guide.

(or 'dangling') bonds present in covalent materials, and lead to gross changes in electronic properties. This is widely exploited in semiconducting device materials, of course. In ionic materials the variation in composition will lead initially to charge imbalance. The way in which this charge imbalance is neutralized within the crystals introduces changes in electronic, optical and magnetic properties, all of which can also be exploited in devices. Less obvious, perhaps, is that mechanical behaviour and chemical reactivity are also strongly affected by the variation in composition. We shall see later how it is possible to manipulate both the composition and properties of these phases to obtain a wide variety of both interesting and useful materials.

In much of the rest of this book we concentrate on the reasons for the composition range of non-stoichiometric compounds and how they are able to accommodate this stoichiometric variability structurally. We will begin by looking at materials in which the cations have a fixed valence, and then progress to materials in which cations can adopt more than one valence state. This will allow us to separate the chemical effects of the composition variation from the electronic changes that occur when variable valence is allowed, although in many cases both effects are of equal importance. Similarly, we will restrict ourselves to simplified structural considerations relying on point defects only, rather akin to those described in Chapter 1. In later chapters, particularly Chapter 9, some of the real ways in which changes of composition are built into inorganic crystals will be described.

4.2 Determination of the composition range of a solid phase

In a practical sense, how do we determine whether a compound is non-stoichiometric? There are a number of techniques that we can use, some of which will be discussed later in this book, but probably the easiest and the one which often gives the first indications that a compound may have a composition range or a non-integral composition is powder x-ray diffraction.

We can illustrate this with reference to the $MgO-Al_2O_3$ system. Suppose that we take pure MgO and mix with it a few percent of pure Al_2O_3 and heat the mixture in air at about 1200 K until reaction is complete. If we then take an x-ray powder photograph, it will show the presence of two phases: MgO, which will be the major component, and a small amount of the phase spinel, $MgAl_2O_4$. This situation is represented on Figure 4.1 by point A. A repetition of the experiment with gradually increasing amounts of Al_2O_3 will yield a very similar result, but the amount of spinel will increase relative to the amount of MgO, until we heat a 1:1 mixture of MgO and Al_2O_3. At this composition only one phase is indicated on the x-ray powder diagram, $MgAl_2O_4$, at point B on Figure 4.1.

A slight increase in the amount of Al_2O_3 in the reaction mixture again yields an x-ray pattern which shows two phases to be present. Now, however, the

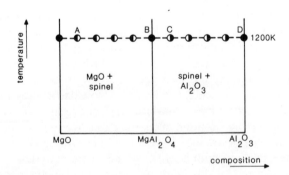

Figure 4.1 Schematic representation of the $MgO-Al_2O_3$ phase diagram at about 1200 K. The half-filled circles represent the fact that preparations will contain two phases, either MgO + spinel, ◑, or spinel + Al_2O_3, ◑, and the filled circles show that only one phase is present. The compositions A, B, C, and D are referred to in the text.

compounds are spinel and Al_2O_3 as indicated by point C in Figure 4.1. This state of affairs continues as we add more Al_2O_3 to the initial mixture, with the amount of spinel decreasing and the amount of Al_2O_3 increasing until we end up with pure Al_2O_3 at point D in Figure 4.1.

Careful preparations reveal the fact that MgO or Al_2O_3 only appear alone on the x-ray films when no additives of the other phase are included in the initial mixtures, and $MgAl_2O_4$ only appears alone at the strictly stoichiometric composition of $1MgO$ plus $1Al_2O_3$, i.e. $MgAl_2O_4$. In addition, over all the composition range studied, the unit cell dimensions of the phases will be unaltered. The results are summarized on the schematic phase diagram shown as Figure 4.1, which indicates that none of the phases appears to have any significant composition range at 1200 K and that they are strictly stoichiometric compounds. The appearance of the phase diagram also gives rise to another name for stoichiometric compounds, viz. *line phases*.

Suppose we repeat this experiment with what at first sight would appear to be a similar pair of oxides, CaO and ZrO_2. When we add a little CaO to ZrO_2 and heat the mixture at about 1200 K we find, on the x-ray power diagram of the resulting product, the presence of a new phase, apart from the ZrO_2 starting material. This situation parallels the behaviour found in the MgO–Al_2O_3 system. As the amount of CaO increases, the amount of the new phase increases, until it finally appears alone in a preparation containing about 13 mole % CaO. This phase, which has a cubic unit cell, is called calcia-stabilized zirconia.

As more CaO is added to the preparations a new phenomenon is noted. No new phase appears on the x-ray diagrams of the material after heating. The calcia-stabilized zirconia phase persists between the approximate limits of 13–28 mole % CaO. A careful examination of the x-ray data also reveals that the unit cell dimensions of this cubic phase change smoothly across the composition range, but there appears to be little else of note recorded on the x-rays. When we finally add more than about 28 mole % CaO to our mixtures we again find the presence of another phase recorded on the x-ray films. This phase increases in amount as we add more CaO, and the system once again seems to be identical in behaviour to the MgO–Al_2O_3 system. The new phase finally appears pure at the $1ZrO_2$:$1CaO$ composition, and has a formula $CaZrO_3$. This behaviour is summarized schematically in Figure 4.2.

The cubic calcia-stabilized zirconia phase is a typical non-stoichiometric compound. It exists over an appreciable composition range, and preparations within the composition range do not have simple formulae. Moreover, the unit cell, although not changing in symmetry, does change its size. In fact a change in unit cell parameters is a very good guide to the existence of a phase range.

The variation in lattice parameter of a non-stoichiometric phase is sometimes found to obey 'Vegard's law', first propounded in 1921, which states that the lattice parameter of a solid solution of two phases with similar structures should be a linear function of the lattice parameters of the

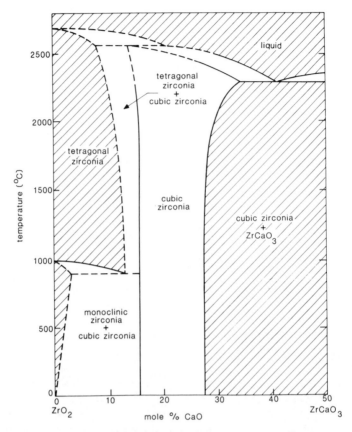

Figure 4.2 Phase diagram of the pseudo-binary CaO–ZrO$_2$ system. The cubic phase occupies the central band in the diagram and is stable to about 2400 °C.

two end members of the composition range. That is, if the lattice parameters of the end members are plotted on a graph of lattice parameter dimension versus composition, the lattice parameters of all the compositions in the intermediate range should lie on the straight line joining these points. In practice Vegard's law rarely holds exactly. Nevertheless, the variation is often regular, and by using this information the composition of a non-stoichiometric phase can often be determined reasonably well without recourse to chemical analysis.

4.3 Non-stoichiometry in compounds containing cations with fixed valence

4.3.1 *Substitution*

If two compounds have the same, or rather similar, structures (as in the case of NiO and MgO or Al$_2$O$_3$ and Cr$_2$O$_3$) then mixed crystals can form with formulae $Ni_xMg_{1-x}O$ or $Al_xCr_{2-x}O_3$. Diffusion in such systems was mentioned in Chapter 2. In these oxides one cation substitutes for another in an otherwise perfect lattice, and although they have broad composition ranges, the metal to oxygen ratio remains unchanged from that of the parent oxides. Because the anion to cation ratio remains fixed, we will consider these materials to be fully stoichiometric, and such compounds will not be considered in any detail here. These mixed phases are often called *solid solutions*, and form most readily when the cations involved are of similar sizes and valence as well as when the end compounds of the solid solution have closely related structures.

Rather than investigating solid solution formation, it is more interesting to consider what happens if we try to substitute cations in a lattice by others which have a different valence. Let us take calcia-stabilized zirconia as an example. When ZrO_2 reacts with CaO, calcia-stabilized zirconia is produced, as was described in the previous section. If we assume that the Ca^{2+} cations occupy positions that are normally filled by Zr^{4+} ions, that is, we have the process of cation substitution, the overall charge neutrality of the crystal will be upset. Clearly, as the Ca^{2+} ions have a lower charge than the Zr^{4+} ions, the crystal will show an overall negative charge if we simply write the formula as $Ca_x^{2+}Zr_{1-x}^{4+}O_2$. Now a few simple experiments will reveal that such a charge build-up does not take place and we must look for a way to overcome this difficulty.

One simple way for the crystal to compensate for the extra negative charge is to arrange for some of the anion sites to be vacant. The number of vacancies on the anion sub-lattice needs to be exactly the same as the number of calcium ions in the structure for exact neutrality. Thus, each Ca^{2+} added to the ZrO_2 produces an oxygen vacancy at the same time. The anion to cation site ratio remains at 2:1, but some of the anion sites are now empty, and the formula of the crystal, if this does indeed take place, becomes $Ca_xZr_{1-x}O_{2-x}$. This need not happen, of course, and we can envisage other ways of achieving charge balance, some of which we will meet in the following section. However, if a material is 'doped' with cations of lower charge, anion vacancies are a common method of achieving charge balance in practice.

Let us now consider the situation where we introduce a cation of higher valence, say by reacting $CaCl_2$ with $NaCl$. Once again consider only substitution, and suppose that Ca^{2+} ions replace Na^{1+} ions in the now impure $NaCl$ crystal. In this case, each Ca^{2+} in the crystal will increase the amount of positive charge in the crystal. One way in which we can balance this is to create a vacancy on the Na^{1+} sub-lattice for each Ca^{2+} incorporated into the structure. Thus it appears that if we dope with an ion of higher valence we may increase the vacancy concentration in the cation sub-lattice when this mechanism holds.

We can see that substitution can produce results rather similar to Schottky defects in the crystal. However, in this case the anion and cation vacancies are no longer balanced, but adjusted to keep the crystal electrically neutral overall.

To prove whether such models do occur in practice we must turn to experimental results. One change in the properties of a crystal which would tend to confirm such models would be an enhancement of the diffusion coefficient of the atoms on the sub-lattice which was believed to contain the extra vacancy population. Thus we also see that doping will allow us to modify the properties of materials, often to useful ends. This aspect will be considered in more detail in the following chapter.

4.3.2 Interpolation

While substitution of one atom type by another of different valence can lead to defects related to Schottky disorder, the presence of interpolated atoms, which are atoms occupying normally empty interstitial positions in the structure, can be considered to be similar to Frenkel defects. As with Frenkel defects, the possibility of finding interstitial atoms in the crystal depends upon the openness of the structure and the size of the impurity atoms. Thus certain obvious size rules must operate, and either the interpolated atoms must be small, or else the host structure very open, with sufficient available space not only for the foreign atoms to be accommodated, but also for them to diffuse into their allocated positions.

Despite this apparent limitation, non-stoichiometric materials which utilize an interstitial mechanism are many and varied. At one extreme are the interstitial alloys, formed when small atoms such as C or N fit into the spaces between larger metal atoms such as Fe. These interstitial phases are extremely important in controlling the hardness and workability of the alloys so produced. At the other extreme we have very open structures which are able to take in molecules. Examples of such compounds are the zeolites, which can accommodate molecules in 'cages' in the structure, and layered materials such as clays which can take in molecules between the layers of which they are composed. In the correct circumstances, quite large molecules can be interpolated between the layers, and these materials are generally referred to as intercalation compounds. In both zeolites and intercalation compounds the surroundings of the incorporated molecules allow for a unique geometrically constrained chemistry to be performed. For this reason both of these latter groups of materials find many uses in fields such as catalysis, but although they are of considerable interest for that reason they do not involve lattice defects to any extent, as the interpolated species are neutral molecules.

In the context of the present volume it is of greater interest to list a few examples from materials which do show marked changes in defect structure. Perhaps one of the most studied oxides from this point of view is the oxygen-rich form of uranium dioxide, UO_{2+x}. The parent phase, UO_2, has the same structure as the mineral fluorite, CaF_2, which is shown in Figure 4.3. In this structure, each uranium atom is at the centre of a cube of eight oxygen atoms. As there are twice as many oxygen atoms as uranium atoms,

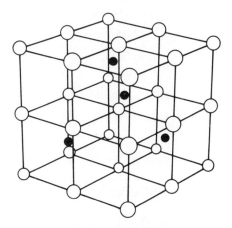

Figure 4.3 The crystal structure of CaF_2, fluorite. The fluorine atoms, drawn as open circles, are in a simple cubic array; the calcium atoms shown as small, filled circles occupy one-half of the cube centres so formed.

half of these cubic sites will be empty in stoichiometric UO_2. It is reasonable, therefore, to assume that the non-stoichiometry in UO_{2+x} is due to interstitial oxygen atoms placed at random in these positions. This explanation of the non-stoichiometry in UO_2 remained in favour for some years but, more recently, improved methods of structural analysis have shown that the interstitial oxygen atoms are not placed at random in the structure at all. In fact they are arranged in local clusters with quite specific geometries, which will be described in more detail in Chapter 9. Despite this fact, we can still regard UO_{2+x} as our first example of a crystal which makes use of interstitials to accommodate a composition range.

The use of interstitials is, in fact, quite common in materials with the fluorite structure. If CaF_2 itself is reacted with LaF_3, YF_3, ThF_4 or similar fluorides, non-stoichiometric phases form. In these, the metal atoms substitute for calcium on the metal ion sub-lattice in a similar way to that described in the case of calcia-stabilized zirconia, but vacancies are not the compensating defect utilized. Charge balance is ensured by the incorporation of F^{-1} ions into the crystals as interstitials which occupy the vacant cube centres. As in the case of UO_2, these are often arranged not at random, but in clusters, as we shall discuss in Chapter 9.

Good examples of non-stoichiometry due to the incorporation of interstitials in layered materials are provided by the layer structure disulphides such as M_xTiS_2, M_xNbS_2 and M_xNbS_2, in which metal atoms (M) occupy positions between sheets of the original disulphide. To illustrate this behaviour, let us look at the titanium sulphides in the composition range between TiS and TiS_2. The structure of TiS_2 is of the CdI_2 type, and is shown in Figure 4.4(a), while the structure of TiS is of the closely related $NiAs$ type shown in Figure 4.4(b). The structures of both TiS and TiS_2 are made up of

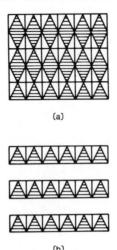

(a)

(b)

Figure 4.4 The structures of (a) TiS_2, and (b) TiS shown as a packing of TiS_6 octahedra. In both structures the sulphur atoms are arranged in a hexagonal stacking. In TiS_2, the octahedral sites between alternate pairs of sulphur atom sheets are filled by titanium atoms, and in TiS all the octahedral sites are filled.

layers of sulphur atoms stacked up in a hexagonal fashion. In TiS all the octahedral sites formed by this stacking are filled, while in TiS_2 all the octahedral sites between every other layer are filled. A range of non-stoichiometric phases can then be envisaged with compositions between TiS and TiS_2 if the vacant octahedral sites in this latter material are filled. This indeed takes place, both with titanium and with other metal atoms. The charge problems which arise in phases such as calcia-stabilized zirconia do not arise in these materials because the outer electrons are delocalized to an appreciable extent, giving the compounds metallic properties. Depending upon the preparation conditions used, the interstitial atoms may be ordered, disordered or partially ordered. If the interstitial atoms are ordered, of course, we no longer have 'defects' present but new phases. In this way a number of intermediate phases in the $TiS-TiS_2$ have been recognized, among which are compounds with the nominal formulae Ti_8S_9, Ti_4S_5, Ti_3S_4, Ti_2S_3 and Ti_6S_8.

The final example of interpolation to cite here is the tungsten 'bronze' Li_xWO_3. The tungsten bronzes are so called because when they were first discovered, by Wohler in 1837, their metallic lustre led him to believe that he had made some new alloys of tungsten rather than new oxides. The lithium tungsten bronze has a composition range from WO_3 to $Li_{0.5}WO_3$ and its structure is quite simply related to that of WO_3. The tungsten trioxide structure is shown in Figure 4.5(a). It consists of a three-dimensional chessboard arrangment of WO_6 octahedra which have quite large cages available for occupancy by foreign atoms or ions. In lithium tungsten bronze the lithium atoms use these sites to form a non-stoichiometric phase, which

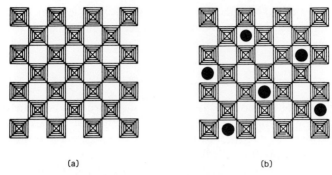

(a)   (b)

Figure 4.5(*a*) The structure of tungsten trioxide, WO_3, which consists of an infinite array of corner-shared WO_6 octahedra. The large cage sites between the octahedra take in interpolated lithium atoms shown as filled circles to form the tungsten bronze Li_xWO_3, as shown in (*b*).

has a structure of the type shown in Figure 4.5(*b*). The phases, like the titanium sulphides noted above, are metallic in nature. The open nature of the tungsten trioxide structure allows the lithium atoms to move readily in and out of the crystals, and as the bronze is dark blue-black in colour, while tungsten trioxide itself is almost colourless in thin sections, these compounds find potential applications in display devices.

4.4 Density and defect type

In section 4.1 the fact that many solids show broad composition ranges was reported, and in section 4.2 some ways in which point defect populations could account for these composition ranges were considered. It is now necessary to look for ways in which the defect structure of a non-stoichiometric compound can be determined. The primary way in which structure is determined is by way of x-ray diffraction, with neutron diffraction and electron microscopy having important roles to play in the case of non-stoichiometric phases. These techniques are sophisticated, and it comes as a surprise to learn that the simple technique of density measurement can give useful clues in this direction. This topic forms the content of the present section, while the structures of non-stoichiometric compounds derived from diffraction techniques are described in Chapter 9.

An x-ray powder photograph yields a measurement of the average unit cell dimensions of a large volume of material, and, for a non-stoichiometric compound, this invariably changes in a regular way across the phase range. In a similar way, the density of a material gives the average amount of matter in a large volume of material, and for a non-stoichiometric phase this also varies across the phase range. These two techniques can be used in conjunction with each other to determine the most likely point defect type in a material. As both techniques are averaging techniques they say nothing about the real organization of the defects, but they do suggest first approximations. We will illustrate the method by reference to two typical examples: iron oxide, with a composition clóse to FeO, and calcia-stabilized zirconia, referred to earlier in this chapter.

4.4.1 Iron monoxide, wüstite, $\approx FeO$

Iron monoxide, often known by its mineral name of wüstite, is found to possess a range of compositions close to $FeO_{1.0}$. X-ray diffraction reveals that the unit cell is cubic and that the structure of wüstite is, to a reasonable approximation, identical to that of rocksalt, NaCl. The lattice parameter, a, is about 0.43 nm and varies as the composition changes. Hence, by means of an experiment it is possible first to obtain an accurate value for the lattice parameter and then to obtain an accurate value for the metal to oxygen ratio by chemical analysis. In this way we can obtain data similar to that listed in Table 4.2. This data shows that there is more oxygen present than iron in the compound. We can consider, as an initial step, two possible point defect models to account for this result.

Model A: In this model we assume that the iron atoms in the crystal are in a perfect array, identical to the Na atoms in NaCl. In this case, to obtain an excess of oxygen, we need interstitial oxygen atoms to be present, as all the normal anion positions, those occupied by Cl in the NaCl structure, will already be full. Now the unit cell of the rocksalt structure contains 4 atoms of Na and 4 atoms of Cl. Hence, in this model, the unit cell must contain 4 atoms of Fe and $4(1 + x)$ atoms of oxygen; that is, the unit cell contents are Fe_4O_{4+4x} and the composition is FeO_{1+x}, where $(1 + x)$ is the figure given in column 1 of Table 4.2.

Model B: On the other hand, we could assume that the oxygen array is perfect and each oxygen atom occupies a site equivalent to that of each Cl atom in the NaCl structure. As we have more oxygen atoms than iron atoms, we must, therefore, have some vacancies in the iron positions. In this case, one unit cell will contain 4 atoms of oxygen and $(4 - 4x)$ atoms of iron. The true formula of $\approx FeO$ now should be written $Fe_{1-x}O$.

It is quite easy to determine which of these suppositions is correct by comparing the real and theoretical density of the material. For example, consider the sample specified in the top line of Table 4.2. The volume v of the

Table 4.2 Experimental data for $\approx FeO^*$

Oxygen: iron ratio	Iron: oxygen ratio	Lattice parameter (nm)
1.058	0.945	0.4301
1.075	0.930	0.4292
1.087	0.920	0.4285
1.099	0.910	0.4282

*The data here and in Table 4.3 are classical data from the paper of E.R. Jette and F. Foote, *J. Chem. Phys.* **1** (1933) 29.

unit cell is given by a^3. In this case the volume is thus $(0.4301 \times 10^{-9})^3 \, \text{m}^3$, so $v = 7.9562 \times 10^{-23} \, \text{cm}^3$.

The mass of a unit cell is readily calculated knowing that the atomic weight of Fe is 55.85 and that of oxygen is 16.00, and that these values correspond to the weight in grams of Avogadro's number, N_A, of atoms. Hence:

Model A: the mass of 1 unit cell is m_A, where

$$m_A = [(4 \times 55.85) + (4 \times 16 \times (1 + x))]/N_A$$

The value of $(1 + x)$, from Table 4.2, column 1, is 1.058, so that

$$m_A = [(4 \times 55.85) + (4 \times 16 \times 1.058)]/N_A \, \text{g}$$

Model B: the mass of one unit cell is m_B, where

$$m_B = [(4 \times (1 - x) \times 55.85) + (4 \times 16)]/N_A$$

and taking the value of $(1 - x)$ form Table 4.2, column 2, we find

$$m_B = [(4 \times 0.945 \times 55.85) + (4 \times 16)]/N_A \, \text{g}$$

The density is, of course, given by the mass of one unit cell divided by the volume:

Model A:

$$[(4 \times 55.85) + (4 \times 16 \times 1.058)]/N_A \times (7.9562 \times 10^{-23})$$
$$= 6.074 \, \text{g cm}^{-3}$$

Model B:

$$[(4 \times 0.945 \times 55.85) + (4 \times 16)]/N_A \times (7.9562 \times 10^{-23})$$
$$= 5.740 \, \text{g cm}^{-3}$$

The difference between the two values is surprisingly large, and is well within the accuracy of density determinations. The value found experimentally is $5.728 \, \text{g cm}^{-3}$, in good accord with the model which assumes vacancies on the iron positions. The indicates that the formula of the $\approx$ FeO phase should be written $Fe_{0.945}O$. Table 4.3 is an expanded version of Table 4.2 and includes the density data. All results are seen to be in good agreement with an experimental formula for $\approx$ FeO of $Fe_{1-x}O$, in which there are vacancies at some of the Fe positions.

Table 4.3 Experimental and theoretical densities of $\approx$ FeO

Oxygen: iron ratio	Iron:oxygen ratio	Lattice parameter (nm)	Observed	Density g cm^{-3} Interstitial oxygen	Iron vacancies
1.058	0.945	0.4301	5.728	6.704	5.740
1.075	0.930	0.4292	5.658	6.136	5.706
1.087	0.920	0.4285	5.624	6.181	5.687
1.099	0.910	0.4282	5.613	6.210	5.652

It is important to remember that this result tells us only that in the unit cell there are vacancies. The arrangement of these vacancies is not indicated at all, and as we shall see in Chapter 9, this arrangement is closely linked to the structure of both FeO and its neighbouring oxide, Fe_3O_4.

4.4.2 Calcia-stabilized zirconia

Now that the principle of using density measurements to clarify the defect structure of crystals has been outlined, we will consider a further example rather more briefly. For this we have chosen calcia-stabilized zirconia. Earlier in this chapter it was noted that materials with the fluorite structure could accommodate a degree of non-stoichiometry either by use of anion vacancies or by way of anion interstitials. In our previous discussion of calcia-stabilized zirconia we suggested that the Ca^{2+} ions occupied sites normally occupied by the Zr^{4+} ion, and to maintain charge balance vacancies were introduced into the anion sub-lattice to yield a formula for the compound of $Zr_{1-x}Ca_xO_{2-x}$. The charge balance requirements meant that one Ca^{2+} ion added to the structure produced one anion vacancy.

We can check this model and confirm that we do have vacancies rather than interstitials in the crystals if we measure the density of the material and compare it with the theoretical density of the fluorite phase. Turning to some experimental results, it has been found that a crystal prepared by heating 85 mole% ZrO_2 with 15 mole% CaO at 1600 °C yielded a material with a cubic unit cell which had a lattice parameter of 0.5560 nm. Remembering our substitution model, we can guess a formula of $Zr_{0.85}Ca_{0.15}O_{1.58}$ for the phase.

The structure of the parent material is cubic, with a unit cell edge of about 0.55 nm. Each unit cell contains four normally occupied cation sites and eight normally occupied anion sites, so that the overall composition of stoichiometric fluorite structure phases is MX_2. In the non-stoichiometric phase we are considering, the mass of the unit cell of this material will be given by

$$m = [4 \times (0.85 \times 91.22) + 4 \times (0.15 \times 40.08) + 8 \times (1.85 \times 16)]/N_A$$
$$= [570.996]/N_A$$

where 91.22 is the atomic weight of Zr, 40.08 is the atomic weight of Ca, 16 is the atomic weight of O and N_A is Avogadro's number. The volume v of the unit cell will be a^3, that is

$$v = (5.560 \times 10^{-8})^3 \, cm^3$$

so that the density is

$$= 5.512 \, g \, cm^{-3}$$

The measured density was found to be $5.477 \, g \, cm^{-3}$. This measure of agreement shows that the substitution plus anion vacancy model is reasonable for this material, but perhaps not perfect.

The experimental data is shown in a more extended form in Figure 4.6. The

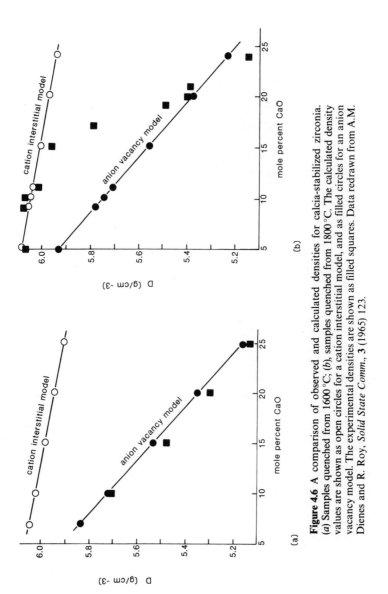

Figure 4.6 A comparison of observed and calculated densities for calcia-stabilized zirconia. (*a*) Samples quenched from 1600 °C; (*b*), samples quenched from 1800 °C. The calculated density values are shown as open circles for a cation interstitial model, and as filled circles for an anion vacancy model. The experimental densities are shown as filled squares. Data redrawn from A.M. Dienes and R. Roy, *Solid State Comm.*, **3** (1965) 123.

calculated values for the density of the samples has been shown for the vacancy model that we have just described and also for a model in which it has been assumed that there are cation interstitials in the crystal instead of anion vacancies. As in the case of wüstite, these densities are sufficiently different to discriminate between the two models. Also shown are the experimentally determined densities for two series of samples, one prepared at 1600 °C and the other at 1800 °C. The results are interesting. It is seen that at 1600 °C the vacancy model fits the data pretty well. In the samples prepared at the higher temperature, however, the situation is more complex. For small amounts of CaO it seems that we have interstitials present. At higher concentrations of CaO, however, the situation is less clear, and it would appear that we pass through a region in which some cells seem to contain interstitials, while in others the substitution mechanism is still employed. Of course, an alternative suggestion is that neither the interstitial nor substitution mechanisms hold at higher temperatures and something different is happening in the structure that density measurements do not allow us to resolve. This aspect will be covered in Chapter 9.

4.5 Defect chemistry

It has already been remarked that point defect populations affect both the physical and chemical properties of materials profoundly. In order to describe these consequences we need a notation for defects that is simple and self-consistent. Indeed, just as the study of chemistry did not enter a period of growth until chemical formulae could be written correctly and chemical reactions expressed in terms of balanced chemical equations, so the study of defect chemistry and physics was hampered until a nomenclature was invented for the purpose of describing the reactions of defects, both with themselves and with the other chemical species, atoms, ions and electrons, in the crystal. The most widely employed system, and the one we shall use in this book, is the Kröger–Vink notation which was designed to account for point defect populations in crystals.

One of the most difficult problems, when working with defects in ionic crystals, is to decide on the charge on the entities of importance. In the Kröger–Vink notation this problem is overcome in the following way. When we add or subtract elements from the crystal, we do so by adding or subtracting electrically neutral atoms and thus avoid making judgements and decisions about chemical bond types. When ionic crystals are involved this requires that we separately add or subtract electrons. To illustrate the implications of this idea we will use the notation to describe some defects in a compound of formula MX, where M is a metal and X an anion. It is simplest to do this by discussing the various types of defect that can occur in such a material, and to commence with uncharged atomic defects.

4.5.1 Atomic defects

Vacancies. When empty lattice sites occur, they are indicated by the symbols V_M and V_X for the metal (M) and non-metal (X) sites respectively. In the notation the subscript M indicates a missing metal atom and X a missing non-metal. If we suppose that the oxide NiO is ionic, V_{Ni} would imply the removal of a Ni^{2+} ion together with two electrons, that is, a neutral Ni atom. Similarly, V_O would indicate a vacancy in the oxygen sub-lattice and implies removal of an O^{2-} ion from the crystal and the subsequent addition of two electrons to the crystal.

Interstitial atoms. When atoms occupy interstitial positions, they are denoted by M_i and X_i for metals and anions respectively. Hence K_i represents an interstitial potassium atom in a crystal.

Impurity atoms. Many materials contain impurity atoms, introduced either on purpose, or as a result of inadequate purification procedures, and it is important to be able to specify the nature of the impurities and where in the crystal they are to be found. This is particularly true for impurities that are deliberately added to control electronic or other properties. In this case the impurity is given its normal chemical symbol and the site occupied is written as a subscript. Thus an Mg atom on a Ni site in NiO would be written as Mg_{Ni}. The same nomenclature is used if an atom in a crystal occupies the wrong site. Thus it is possible for M atoms to be on X sites, written as M_x, or X atoms to be on M sites, written as X_M. A potassium atom on a bromine site in KBr would be written as K_{Br}, for example.

Associated defects. As we will see in following sections, it is also possible for one or more lattice defects to associate with one another, that is, to cluster together. These are indicated by enclosing the components of such a cluster in parenthesis. As an example, $(V_M V_X)$ would represent a Schottky defect in which the two vacancies were associated as a vacancy pair.

The foregoing notes show that the nomenclature uses a straightforward system of description. It is seen that the normal symbol for a chemical element represents the species involved, and the subscript represents the position of the atom in the structure. Apart from the symbols for chemical elements, we have used 'V' to mean vacancy and 'i' to mean interstitial. The symbol V is also the chemical symbol for the element vanadium, of course. Where confusion may occur, the symbol for a vacancy is written 'Va'.

4.5.2 Charges on defects

Electrons and electron holes. The charged defects that most readily come to mind are electrons. Some fractions of the electrons will be free to move through the crystal. These are denoted by the symbol e'. The superscript $'$ represents the negative charge on the electron. Although electrons are the only charged sub-atomic particles to exist in the structure it often simplifies matters to think about the sites where electrons are missing. This is analogous to thinking about vacancies instead of atoms. In the case of these 'electron vacancies' we use the symbol $h^{\cdot}$ to denote the defect, which is called an electron hole, or, more commonly, simply a 'hole'. Each hole will bear a positive charge of $+1$, which is represented by the superscript $^{\cdot}$.

Charges on defects. Besides the electrons and holes just mentioned, the atomic defects that we have described above can also carry a charge. In ionic crystals, in fact, this may be considered to be the normal state of affairs. The Kröger–Vink notation bypasses the problem of deciding on the real charges on defects by considering only effective charges on defects. The *effective charge* is the charge that the defect has *with respect to the normal crystal lattice*. To illustrate this concept, let us consider the situation in an ionic material such as NaCl, which we will suppose to be made up of the charged ions Na^+ and Cl^-.

If we then have a vacancy in the NaCl structure at a sodium position V_{Na}, what will the effective charge on this defect be? To understand this, you must imagine yourself as 'diffusing' through the NaCl structure. Each time an Na^{1+} ion is encountered, a region of positive charge will be experienced. If, then, we meet a vacancy instead of a normal ion, this will seem not to be positive at all. Relative to the situation normally met with at the site we will encounter a region which has an effective negative charge, that is, a charge relative to that normally encountered at that position equivalent to -1. In order to distinguish effective charges from real charges, the superscript $'$ is used for each unit of negative charge and the superscript $^{\cdot}$ is used for each unit of positive charge. Hence a 'normal' vacancy at a sodium site in NaCl would be written as V'_{Na}, which corresponds to a missing Na^{1+} ion. Similarly, a 'normal' vacancy at a chlorine site would seem to be positively charged relative to the normal situation in the crystal. Hence the vacancy has an effective charge of $+1$, which would be written $V^{\cdot}_{Cl}$.

With each of the other defect symbols V_M, V_X, M_i, M_X and associated defects such as $(V_M V_X)$ an effective charge relative to the host lattice is also possible. Thus $Zn^{2\cdot}_i$ would indicate a Zn^{2+} ion at an interstitial site which is normally unoccupied and hence without any pre-existing charge. In such a case, all the charge on the Zn^{2+} ion is experienced as we move through the lattice, and hence the presence of two units of effective charge is recorded in the symbol, viz. $2^{\cdot}$. Similarly, substitution of a divalent ion such as Ca^{2+} for monovalent Na on a sodium site gives a local electronic charge augmented by one extra positive charge which is then represented as $Ca^{\cdot}_{Na}$.

Suppose now a sodium ion in NaCl, represented by Na_{Na}, is substituted by a potassium ion, represented by K_{Na}. Clearly the defect will have no effective charge, as, to anyone moving through the crystal, the charge felt on

Table 4.4 Summary of the Kröger–Vink notation*

Type of defect	Notation	Type of defect	Notation
Non-metal vacancy at non-metal site	V_X	Charged interstitial metal	$M_i^{\cdot}$
Metal vacancy at metal site	V_M	Charged interstitial non-metal	X_i'
Interstitial metal	M_i	Free electrons**	e'
Interstitial non-metal	X_i		
Foreign metal, (A), at metal sites	A_M	Free positive holes**	$h^{\cdot}$
Foreign non-metal, (Y), at non-metal site	Y_X		
Neutral vacancies	$V_X^x \quad V_M^x$		
Positively charged non-metal vacancies	$V_X^{\cdot}$		
Negatively charged metal vacancies	V_M'		

*The definitive definitions of this nomenclature and further examples are to be found in the IUPAC *Red Book on the Nomenclature of Inorganic Chemistry*, Chapter I-6.
**Concentrations of these defects are designated by n and p, respectively.

encountering the K ion is the same as that experienced on encountering a normal Na ion. This defect is therefore neutral in terms of effective charge. This is written as K_{Na}^x when the effective charge situation needs to be specified, the superscript x representing an effectively neutral charge.

It is therefore seen that the idea of the charge on the defect is separated from the chemical entity which makes up the defect. Real charges are represented by n^+ and n^-, while effective charges are represented by n', and $n^{\cdot}$, or x. It is for this reason that the charges on electrons and electron holes mentioned above were written as $'$ and $^{\cdot}$, as these charges are also of importance only relative to the surrounding crystal lattice.

The main features of the Kröger–Vink notation are summarized in Table 4.4.

4.5.3 *Reaction equations*

There are many instances where we have to consider reactions which cannot be expressed within the normal chemical nomenclature. For example, if an impurity is doped into a crystal it can have profound effects upon the physical and chemical properties of the substance because of the defects necessarily introduced. However, defects do not occur in the balance of reactants expressed in the traditional equations and so these important effects are lost to the chemical accounting system that the equations represent. If defects can be incorporated into normal chemical equations, it will not only allow us to keep a strict account of these important entities but also to apply chemical thermodynamics and other techniques of handling chemical energy exchange to the reactions. We can therefore build up a *defect chemistry*, in which the defects play a role analogous to that of the chemical atoms themselves. The Kröger–Vink notation allows this to be done, provided the normal rules which apply to balanced chemical equations are preserved. As the rules are slightly different to those of elementary chemistry they are set out here.

(i) The number of metal atom sites must always be in the correct proportion to the number of non-metal atom sites in the crystal. Thus, in MgO we must always have equal numbers of both types of position; in TiO_2, there must always be twice as many anion sites as cation sites, and in a compound M_aX_b, there must be a metal atom sites for every b non-metal atom sites. As long as this proportion is maintained, the total number of sites can vary, as this simply corresponds to more or less substance present. If the crystal contains vacancies these must also be counted in the total number of sites, of course, as each vacancy can be considered to occupy a site just as legally as an atom. Interstitial atoms do not occupy normal sites and so do not count when this rule is being applied.

(ii) As the equations to be written are chemical equations, the total number of atoms on one side of the equation must balance the total number of atoms on the other side. Remember that the subscripts and superscripts are labels describing charges and sites, and are not counted in evaluating the atom balance.

(iii) The crystal must always be electrically neutral. This means not only that the total charge on one side of the equation must be equal to the total charge on the other side, but also that the sum of the charges on each side of the equation must equal zero. In this assessment, both effective and real charges must be counted if both sorts are present.

(iv) When crystals react, only neutral atoms are involved. After reaction, neutral atoms can dissociate into charged species if this is thought to represent the real situation in the crystal.

As an example to illustrate exactly how these rules work, and to show that their application is not difficult, let us consider the reactions that can occur when crystals of ZrO_2 are reacted with CaO to produce a crystal of calcia-stabilized zirconia. This sort of situation cannot be treated by normal chemical equations, but it is clear that such reactions do take place and are important.

In ZrO_2 there are twice as many anion sites as there are cation sites. Let us suppose, as we have in the past, that the Ca atoms will be located on normal cation sites. If each Ca atom in the structure is located on a cation site we have created one new cation site per Ca atom. In order to comply with rule (i) we must therefore create two anion sites per Ca atom. These are considered to be vacant at the start. However, as we have to locate an oxygen atom in the crystal as well, it is reasonable to place it in one of these sites. The other site remains vacant. We also have noted that the reactions are to be carried out using neutral atoms, so as to avoid mistakes over the allocation of charges to reacting species. The reaction equation is then

$$CaO\,(c) \xrightarrow{ZrO_2} Ca_{Zr}^{4'} + V_O^{2\cdot} + O_O^{2\cdot} \qquad (4.1)$$

This means that in the structure of ZrO_2 we now have Ca atoms and O atoms on sites normally occupied by metal and non-metal species. As the Ca atom is taken as being neutral, the effective charge at the site will be $4'$ with respect to the charge encountered when a normal Zr ion is met with. Similarly, the oxygen atom will be neutral and so the effective charge upon the oxygen site which is occupied will be $2\cdot$, and that upon the oxygen site which is vacant will equally be $2\cdot$. Note that the equation conserves mass balance, electrical charge balance and site numbers, in accordance with the rules given above.

Now we may feel that this is unreasonable and prefer to think of ZrO_2 and CaO as ionic compounds, so that ions should occupy the sites, not neutral atoms. This gives us an alternative and more realistic process which we can write as

$$CaO(c) \xrightarrow{ZrO_2} Ca_{Zr}^{2'} + V_O^{2\cdot} + O_O \qquad (4.2)$$

This equation also conserves mass balance, electrical charge balance and site numbers, as indeed it must. We notice that the effective charge on the Ca ion is now $2'$ as the normal charge at a cation position is due to the presence of Zr^{4+} ion, and so, with respect to the normal situation, the presence of the Ca ion leads to an effective decrease in the charge encountered at the site in question by two units. Similarly, the oxygen ion occupies a normal oxygen ion site and we have no difference from that normally encountered in ZrO_2, and so the effective charge for these ions is zero.

It may be argued that the Ca ions do not occupy Zr sites, but prefer interstitial positions. The Ca ions are then easy to deal with, as they do not affect the site number in the ZrO_2 matrix. Each oxygen atom can again be assumed to occupy an anion site. Now in this case the site conservation rule applies, and for each fresh anion site created we must create one half of a new cation site. This simply means that every two oxygen atoms incorporated into the crystal generate one new cation position. As the Ca ions do not make use of these positions they remain empty. Once again, taking the atomic entities to be ions rather than neutral atoms, we can write the formation equation as

$$2CaO(c) \xrightarrow{ZrO_2} 2Ca_i^{2\cdot} + 2O_O + V_{Zr}^{4'} \qquad (4.3)$$

So far we have written down three equations which could apply to the reaction of CaO with ZrO_2 to form calcia-stabilized zirconia. All of them are correct in a chemical sense. To decide which of them, if any, represents the true situation in the material, experimental evidence, such as that derived from density measurements, must be sought.

4.6 Point defect interactions

We have seen that defects in a crystal can carry effective charges, and because of this we would expect the defects to interact with each other quite strongly. For one reason or another it may be important to try to calculate these energies, and it is easy to imagine that such calculations will be fairly difficult. Before we embark on this task it would therefore be worthwhile to see if we can gain some approximate feeling for the magnitude of these energy terms. If they turn out to be very small it may not be worthwhile to undertake the more exact calculations at all. As we are thinking about the interaction of charged defects, perhaps the place to start is with simple electrostatic theory. This gives the energy of interaction of two unit charges (sometimes expressed as the work needed to separate them) as

$$E_{electro} = e^2/4\pi\varepsilon_0 r \qquad (4.4)$$

where each of the charges has a magnitude of e and we assume that the charges have opposite signs, and so attract each other, r is the separation of the charges and ε_0 is the permittivity of vacuum.

If we apply this formula to defects in a crystal, and again assume that the defects are oppositely charged, so that they attract each other, the energy term will be roughly equivalent to the enthalpy of formation of a defect pair, ΔH_p. In order to allow for the crystal structure itself, which will modify the interaction energy considerably, we make the assumption that the force of attraction is simply 'diluted' in the crystal by an amount equal to its dielectric constant. The modified formula is then

$$\Delta H_p = (z_1 e)(z_2 e)/4\pi\varepsilon_0\varepsilon r \qquad (4.5)$$

where ΔH_p is the enthalpy of interaction, z_1 and z_2 are the effective charges on the defects, ε is the static dielectric constant of the crystal and the other symbols have the same meaning as in equation (4.4).

Certainly this theory is not very refined, but it should be good enough to tell us whether association of defects is likely to occur or not. Let us consider, as an example, a Schottky defect, consisting of a cation vacancy and an anion vacancy, in a crystal of a monovalent metal MX with the rocksalt structure. As we have seen, these vacancies will have effective charges of $+e$ and $-e$. Their interaction will be greatest when they are closest to each other, that is, when they occupy neighbouring sites in the crystal. The separation of these sites is about 3×10^{-10} m. A typical value for the dielectric constant of a rocksalt structure crystal is about 10. The value of ε_0 is given by $8.854 \times 10^{12}\,\mathrm{F\,m^{-1}}$ and the electronic charge by $1.6022 \times 10^{-19}\,\mathrm{C}$, so the interaction energy, which is attractive, is given by

$$\Delta H_p = (1.6022 \times 10^{-19})^2/4 \times 8.854 \times 10^{-12} \times 10 \times 3 \times 10^{-10}$$

$$= 7.691 \times 10^{-20} \qquad (4.6)$$

Before we can say whether this is a large or a small interaction enthalpy, we must determine the units. The value we found in equation (4.6) has units

$$\Delta H_p = \mathrm{C^2/F\,m^{-1}\,m} \qquad (4.7)$$

$$= \mathrm{C^2/F} \qquad (4.8)$$

which is not a familiar energy unit. The matter can be resolved by converting the units above into their more fundamental values, thus:

$$\mathrm{C = A\,s,\ F = C\,V^{-1},\ V = W\,A^{-1}\ and\ W = J\,s^{-1}}$$

where C = coulomb, A = ampere, F = farad, V = volt, W = watt and J = joule. Making the appropriate substitutions,

$$\mathrm{C^2/F = C\,V = A\,s\,J\,s^{-1}\,A^{-1} = J} \qquad (4.9)$$

Hence we see that the value is, in fact, in joules, so that

$$\Delta H_p = 7.69 \times 10^{-20}\,\mathrm{J} \qquad (4.10)$$

The figure we have calculated is the interaction energy for one pair of vacancies only. To obtain the molar quantity, we multiply H_p above by Avogadro's number, N_A, to yield

$$\Delta H_p = 46.3\,\mathrm{kJ\,mol^{-1}} \qquad (4.11)$$

How does this compare with other energy terms, such as the energy of formation of Schottky defects? Typical values are given in Chapter 1. The value calculated is very similar in magnitude to the values quoted for Schottky defect formation energies and so we would expect that a reasonable proportion of the vacancies would be associated into pairs.

We can actually make an assessment of the fraction of defects in a crystal which are associated, using either the rough interaction energies calculated above or more accurate theoretical values in the following way.

The Boltzmann equation tells us that if we have two energy states separated by an energy difference ΔH, the fraction of the population in the upper state, f, is given by

$$f = \exp(-\Delta H/kT) \qquad (4.12)$$

where k is Boltzmann's constant and T the temperature. We can use this for our purposes, and to provide an example, let us return to the case discussed above, that of vacancy pair association in an NaCl-type material.

The number of Schottky defects will be n_s, so that we will have n_s cation vacancies and n_s anion vacancies in the crystal. If we take the interaction energy to be 7.69×10^{-20} J, as we calculated, the fraction of vacancies associated will be given by f, where

$$f = \exp\left[(-7.691 \times 10^{-20})/(1.38066 \times 10^{-23} \times 10^3)\right]$$

where we have taken a value of 1.38066×10^{-23} J K for Boltzmann's constant, k, and a temperature of 1000 K. Hence

$$f = 0.0038 \tag{4.13}$$

That is, about 4 defects in every 1000 will be associated into pairs at 1000 K. As we already know how to estimate the number of Schottky defects in a crystal, it is therefore possible to find the total number of vacancies that are associated in pairs. Obviously we can use similar reasoning to that above for other defect types.

We can conclude this section by observing that although the estimates of interaction energy given here are approximate, they do suggest that a fair number of defects will not exist simply as isolated 'point defects', but associated with other point defects in defect clusters. This conclusion has been borne out in recent years by realistic calculations which can be made using computers. These show that much of the interaction energy between point defects arises when the atoms in the crystal close to the defects move slightly to adjust to the new configuration of the defects themselves. This process is known as relaxation, and the relaxation energy so produced makes a major contribution to the overall energy of defect clusters. Experimental work has also supported these theoretical conclusions, and some relevant results are presented in Chapter 9.

4.7 Supplementary reading

Unfortunately there are very few books which cover the material in this chapter at an introductory level. Probably the most complete account of point defect chemistry, and an explanation of the Kröger–Vink notation, is to be found in F.A. Kröger, *The Chemistry of Imperfect Crystals*, 2nd edn., North-Holland, Amsterdam (1974).

An account of the defect chemistry of oxides with much experimental data is P. Kofstad, *Nonstoichiometry, Diffusion and Electrical Conductivity in Binary Metal Oxides*, Wiley-Interscience, New York (1972).

A brief but interesting account of defect chemistry is given in W.J. Moore, *Seven Solid States*, Benjamin, New York (1967), Chapters 1 and 5.

Two very good review articles on the subject matter of this Chapter are J.S. Anderson, in *Chemistry of the Solid State*, ed. C.N.R. Rao, Marcel Dekker, New York (1974); D.J.M. Bevan, Chapter 49 in *Comprehensive Inorganic Chemistry*, Vol. 4, ed. A.F. Trotman-Dickenson, Pergamon, Oxford (1973).

5 Some applications: galvanic cells and sensors using non-stoichiometric compounds

5.1 Introduction

We have seen that diffusion often plays a controlling role in chemical reactions between solids, and that bulk diffusion itself is dominated by the point defect populations to be found in the structure under consideration. Unlike stoichiometric phases, non-stoichiometric compounds are able to vary their defect populations by way of changes in their compositions. In this chapter we try to link these various aspects to show how our understanding of non-stoichiometry will allow us to overcome the normal limitations of diffusion in stoichiometric solids which are due to the inherently low populations of native point defects present. This will allow us to create some useful device materials and understand the way in which devices based on such materials function.

As in Chapter 4, we will limit ourselves to considering principally those materials which contain ions of fixed valence. These materials are insulators, and show no electronic conductivity, which, as we shall see, makes them especially useful as electrolytes in batteries or electromotive cells; but other uses, particularly as gas detectors and sensors, are also important. Our emphasis will be on the way in which the defect populations can be manipulated to enhance the desired properties of the material rather than on cell technology. Firstly, we look at some of the guidelines that we can use to manipulate these populations, and then briefly at the principles which underlie the operation of galvanic cells, before discussing some examples.

To illustrate the principles involved we will concentrate on two broad groups of structures. Firstly, we can add impurities to a crystal so as to produce a non-stoichiometric phase with very high 'point defect' concentrations. Calcia-stabilized zirconia fits this description well. Secondly, we can introduce into the solid large numbers of interstitial ions which will serve to conduct electricity. As typical compact solids do not usually accept large numbers of such ions, we must find particular structures to use. Layer structures are particularly useful in this respect. In this case the interstitial atoms are added between layers of normal structure to form sandwich-like products. Looked at another way, in this class of compounds the point defect population problem is overcome by avoiding the solid matrix and using it only

99

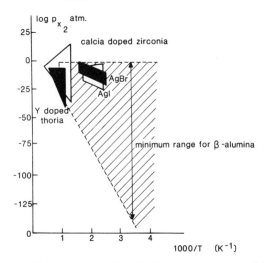

Figure 5.1 Electrolytic conduction domains for several common solid electrolytes.

as a support for the layers of conducting defects. The compound β-alumina typifies this type of material.

The resulting materials have very high ionic conductivities, similar to those of liquid electrolytes. Such materials are sometimes called *super-ionic conductors*, but the term *fast ion conductors* is to be preferred to avoid confusion with metallic superconductors, which transport electrons by a quite different mechanism. Figure 5.1 shows the domains of electrolytic conduction for the principal materials discussed in this chapter, together with that of the silver halides AgI and AgBr mentioned in Chapter 1.

5.2 Galvanic cells

A galvanic cell is an arrangement which enables the energy of a chemical reaction to be utilized directly, often as a portable supply of electricity. *Primary cells* are those that can only be used once, while those cells which are rechargeable are known as *secondary storage cells*. In both these types of cell the materials which are involved in the reaction form the *electrodes*, and the reaction takes place by the passage of ions through an *electrolyte* which separates the electrodes from each other.

The formation of ions during the chemical reaction taking place in the cell involves the transfer of electrons to or from the electrodes. In a galvanic cell these are not allowed to pass through the electrolyte but must pass through an external circuit, driven by a potential difference which is created between the electrodes. It is this electron current which is used to do useful work. In a secondary cell the normal reaction can be driven backwards by overcoming the potential difference between the electrodes by a greater and reversed potential. This has the effect of forcing the electrons and ions to retrace the paths that they take during normal operation.

If the electron flow through the external circuit is stopped, the potential difference between the two electrodes prevents further reaction and chemical equilibrium is reached. Usually when this happens the reaction within the cell is in reversible equilibrium, and the potential difference is very simply related to the free energy of the chemical reaction which is taking place, called the *cell reaction*, by the equation

$$G = -zEF \qquad (5.1)$$

Here G is the Gibbs' free energy change involved in the cell reaction, z is the number of electrons transferred (via the external circuit) in the cell reaction, E is the measured open cell emf, and F is the Faraday constant.

If the electrode components are present at unit thermodynamic activity, and the resistence of the external circuit is very high, we obtain the standard emf of the cell, which yields the standard Gibbs' free energy change for the reaction, thus:

$$G^0 = -zE^0F \qquad (5.2)$$

If a small amount of electronic conduction is also able to take place through the electrolyte, the cell never gives a true open circuit emf as the cell reaction can proceed at a speed which is related to the passage of electrons within the cell itself rather than through the outside circuit. In this case the measured emf is less than the open circuit emf, and equations (5.1) and (5.2) above need to be modified somewhat. For the purposes of our discussion here we do not need to consider this further, but details can be found in the supplementary reading list (section 5.6). In the sections that follow we show how the simple relationship between the free energy of the cell reaction and the emf developed can be exploited.

5.3 Calcia-stabilized zirconia and related fluorite structure oxides

5.3.1 *The structure of calcia-stabilized zirconia*

The stabilized zirconia group of oxides is widely used in solid state electrochemical systems. They have nominal formulae $CaO.x\,AO_2$, where A can be typically Zr, Hf and Th, or $M_2O_3.xZrO_2$, where M can be typically La, Sm, Y, Yb or Sc, and structures closely related to the fluorite type illustrated in the previous chapter. These materials differ quite significantly from most other fast ion conductors in that they have oxygen ion transport numbers with values very close to 1.0. In order to understand how these oxides are able to conduct large oxygen ions with facility we will first consider their defect structure.

The correct composition of a fluorite structure oxide would be MO_2.

Taking calcia-stabilized zirconia as an example, we know that addition of CaO drops the metal to oxygen ratio to below 2.0, and the formula of the oxide becomes $Ca_xZr_{1-x}O_{2-x}$. As we described in sections 4.2.1 and 4.3, the likely structure for calcia-stabilized zirconia fabricated at temperatures of about 1600 °C is one in which the Ca^{2+} ions are found on sites normally occupied by Zr^{4+} ions and we have compensating vacancies on the oxygen sites. For each Ca^{2+} ion inserted into the structure, we must create one anion vacancy. Hence an oxide containing 20 mole % CaO will have 20 mole % oxygen vacancies in the structure. The result of this enormous defect population is to greatly increase the diffusion coefficient of oxygen to the extent that the material becomes a very fast oxygen-ion conductor with an anion transport number close to 1.0.

Note that this property of calcia-stabilized zirconia arises solely from the defect population present, rather than the constituent ions that actually make up the material. Hence materials with related structures should behave in a similar way. This is so. Thus HfO_2, ThO_2 and CeO_2, when reacted with calcia all form structures which contain large numbers of anion vacancies. Similarly, if we combine ZrO_2 with La_2O_3, Sm_2O_3, Y_2O_3 or Sc_2O_3 we again produce similar materials. The population of oxygen vacancies will be rather less per substituted ion in these latter compounds, as the valence of the foreign cations is $3+$ rather than $2+$ as with CaO, leading to a generalized formula of $M^{3+}Zr_{1-x}O_{2-1/2x}$. All of these phases do, nevertheless, retain the property of high oxygen ion conductivity to the virtual exclusion of other ionic or electronic transport processes.

5.3.2 Oxygen sensors

One of the most interesting applications of calcia-stabilized zirconia is as an oxygen meter or *sensor*. This ability arises in the following way. Because the solid is able to transport oxygen ions, when a slab of calcia-stabilized zirconia separates two regions which contain oxygen gas at different pressures, oxygen ions will migrate from one side to the other. This migration will produce a potential, of course, because the ions are charged. Measurement of this potential will give a measure of the oxygen pressure difference, allowing the cell to be used as an oxygen meter. Moreover, as the calcia-stabilized zirconia is sensitive to the passage of oxygen ions alone it can be used to indicate when

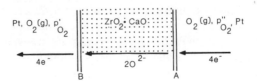

Figure 5.2 Schematic illustration of the processes taking place in a calcia-stabilized zirconia oxygen sensor.

oxygen is present, and thus act as an oxygen sensor. For both of these purposes the calcia-stabilized zirconia forms the electrolyte in a galvanic cell. The cell arrangement is drawn schematically in Figure. 5.2.

In order to understand how we can determine the oxygen pressure on one side of the zirconia with respect to that on the other side, we need to consider the reactions which take place at each side of the electrolyte. At the side marked A in Figure 5.2 we have the reaction

$$O_2(g),[p''_{O_2}] + 4e^- [\leftarrow] \rightarrow 2O^{2-} [\leftarrow] \tag{5.3}$$

At B we have the reaction

$$2O^{2-} [\leftarrow] \rightarrow O_2(g),[p'_{O_2}] + 4e^- [\leftarrow] \tag{5.4}$$

where the arrows in square brackets show the direction of electron or ion flow with respect to the diagram. The relevant oxygen pressures on each side of the zirconia are also noted in these equations. The overall cell reaction will be given by the sum of equations (5.3) and (5.4), which is

$$O_2(g), [p''_{O_2}] \rightarrow O_2(g), [p'_{O_2}] \qquad \Delta G_1 \tag{5.5}$$

By reference to section 5.2 we see that the free energy change of this reaction, ΔG_1, will be given by

$$\Delta G_1 = -4EF \tag{5.6}$$

Chemical thermodynamics tells us that ΔG_1 is simply related to the chemical potentials of the oxygen gas on each side of the calcia-stabilized zirconia in the following way:

$$\Delta G_1 = \mu'_{O_2} - \mu''_{O_2} = RT\ln[p'_{O_2}/p''_{O_2}] \tag{5.7}$$

Equating (5.6) and (5.7) gives equation (5.8):

$$p'_{O_2} = p''_{O_2} \exp[-4EF/RT] \tag{5.8}$$

We can therefore use the cell directly as an oxygen meter if p''_{O_2} is a standard pressure, such as one atmosphere of oxygen, or else the pressure of oxygen in air, which is approximately 0.21 atmosphere. We can also note that in using this equation the only units of relevence are those of oxygen pressure, as the exponential term is dimensionless.

It is quite easy to modify this cell so as to use it to measure the concentration

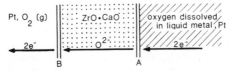

Figure 5.3 Schematic illustration of the processes taking place in a calcia-stabilized zirconia cell used to measure the oxygen content of a liquid metal.

of oxygen in, say, a liquid metal. The schematic cell arrangement is shown in Figure 5.3. To understand how the information that we require is obtained we consider the reactions on each side of the electrolyte, as we did before. At side A we have the reaction

$$O \text{ [dissolved in metal]} + 2e^-[\leftarrow] \rightarrow O^{2-}[\leftarrow] \qquad (5.9)$$

At side B we have

$$O^{2-}[\leftarrow] \rightarrow \tfrac{1}{2}O_2 + 2e^-[\leftarrow] \qquad (5.10)$$

Adding (5.9) and (5.10) gives the overall cell reaction, equation (5.11):

$$O \text{ [dissolved in metal]} \rightarrow \tfrac{1}{2}O_2 \qquad \Delta G_2 \qquad (5.11)$$

ΔG_2, the free energy change for reaction (5.11), is equal to the difference in the chemical potentials of the oxygen gas on each side of the stabilized zirconia, and we can write the following pair of equations:

$$\Delta G_2 = -2EF \qquad (5.12)$$

$$\Delta G_2 = \mu_{O_2}(g) = \mu_O(\text{dissolved}) \qquad (5.13)$$

Replacing the chemical potential terms we find

$$-2EF = RT \ln(a_O^2/p_{O_2}) \qquad (5.14)$$

where a_O is the activity of oxygen in the liquid metal, which is, in dilute

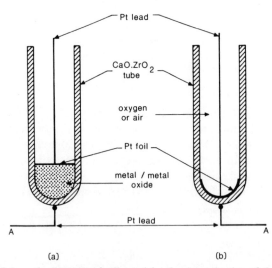

(a)   (b)

Figure 5.4 Schematic diagrams of cells used for the determination of the oxygen content of a liquid or vapour. In (a) a metal/metal oxide couple such as Ni/NiO is used as a standard on one side of the cell, and in (b) oxygen gas at a standard pressure is used. If the aim is to determine the concentration of oxygen dissolved in a liquid metal, the zirconia tube can simply be immersed in the molten metal.

solutions, equivalent to the oxygen concentration in the metal. If we take p_{O_2} as one atmosphere we can simplify the last equation to write

$$-2EF = RT \ln a_O^2 \qquad (5.15)$$

Note that these equations consider the oxygen to be dissociated in solution. If this is not so, and the oxygen exists as molecules of O_2, we will have to modify the equations somewhat to give

$$-4EF = RT \ln a_O \qquad (5.16)$$

The simplest way to make such a meter in practice is to take a calcia-stabilized zirconia tube which is closed at one end, as shown in Figure 5.4. The tube is simply dipped into liquid metal and the equilibrium potential gives a measure of the oxygen activity, and hence oxygen concentration, in the liquid metal.

There are numerous other industrial applications of stabilized-zirconia oxygen meters to be found, as such devices are particularly suitable for remote sensing of oxygen gas in inhospitable environments such as flues and furnaces. Examples of such uses will be found in the Supplementary Reading section at the end of this chapter.

5.3.3 Free energy meters

As the potential developed across a calcia-stabilized zirconia electrolyte is simply related to the free energy of the reactions taking place in the surrounding cell, such cells can be used directly as free energy meters. One of the simplest cells we can envisage is one in which one electrode is oxygen gas at one atmosphere pressure, as has just been described. The cell can measure, for example, the free energy of dissociation of a metal oxide or its equivalent, the oxygen pressure over a metal/metal oxide system. The cell is represented diagrammatically in Figure 5.5, and we have chosen to illustrate its use for determination of the free energy of formation of nickel oxide.

At side A of the electrolyte the reaction taking place is

$$\tfrac{1}{2}O_2(g) + 2e^-[\leftarrow] \rightarrow O^{2-}[\leftarrow] \qquad (5.17)$$

At side B the reaction taking place is

$$Ni + O^{2-}[\leftarrow] \rightarrow NiO + 2e^-[\leftarrow] \qquad (5.18)$$

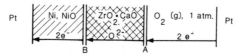

Figure 5.5 Schematic illustration of the processes taking place in a calcia-stabilized zirconia cell used to measure the free energy of formation of NiO.

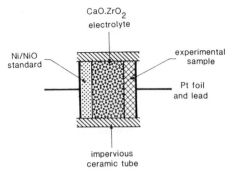

Figure 5.6 Arrangement of the components of a cell used to determine the free energy of formation of metal oxides.

The overall cell reaction is found by adding equations (5.17) and (5.18) to yield equation (5.19), thus:

$$Ni + \tfrac{1}{2}O_2 \rightarrow NiO \qquad \Delta G_3 \qquad\qquad (5.19)$$

The free energy of this reaction is ΔG_3. If the oxygen gas on the reference side of the cell is at one atmosphere pressure, the free energy change measured, ΔG_3, will be equal to the standard free energy of formation of NiO, G^0_{NiO}, which will be given by

$$G^0_{NiO} = -2E^0 F \qquad\qquad (5.20)$$

This sort of cell can also be used to measure the partial oxygen pressure over the Ni/NiO mixture, p_{O_2}, as this is simply given by

$$G^0_{NiO} = -RT \ln p_{O_2} \qquad\qquad (5.21)$$

If the free energy of formation of NiO is known from other measurements we can then use this cell as an oxygen meter, as the cell voltage will be related to the pressure of the oxygen gas at the free electrode, as described above. This application is shown in Figure 5.4. In a similar way, replacement of the oxygen electrode with other metal–metal oxide mixtures allows their free energy of formation to be determined. The cell arrangement used is shown in Figure 5.6 and the schematic diagram for the measurement of the free energy of formation of $\approx$ FeO is shown in Figure 5.7.

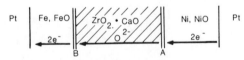

Figure 5.7 Schematic illustration of the processes taking place in a calcia-stabilized zirconia cell used to measure the free energy of formation of $\approx$ FeO.

Referring to Figure 5.7, the reaction taking place at A is

$$2e^-[\leftarrow] + NiO \rightarrow Ni + O^{2-}[\leftarrow] \tag{5.22}$$

The reaction taking place at B is

$$Fe + O^{2-}[\leftarrow] \rightarrow FeO + 2e^-[\leftarrow] \tag{5.23}$$

The overall cell reaction is obtained by adding (5.25) and (5.26), to give

$$Fe + NiO \rightarrow FeO + Ni \qquad \Delta G_4 \tag{5.24}$$

The Gibbs' free energy of the cell reaction is then given by

$$G_4^0 = -2EF = G_{FeO}^0 - G_{NiO}^0 \tag{5.25}$$

That is, the emf simply measures the difference in the free energies of the electrode materials, NiO and FeO. As we are taking NiO to be a standard, we can write

$$G_{FeO}^0 = -2EF + G_{NiO}^0 \tag{5.26}$$

Now equations (5.22) to (5.26) assume that FeO has a fixed composition, which is not true, as we saw in the previous chapter. We can take this into account and use the cell to determine the free energy of the $\approx$ FeO phase as a function of composition in the following way. The NiO component of the cell is replaced by $\approx$ FeO of one composition, say FeO_{1+x_1}, the other component being of a different composition, say FeO_{1+x_2}. The cell reaction involves the transport of one mole of O_2 per 4 faradays of electricity passed between the two sides of the cell. The measured emf is given by

$$E = 1/4[\Delta G(FeO_{1+x_1}) - \Delta G(FeO_{1+x_2})]$$

$$= 1/4\{RT \ln [p_{O_2}(FeO_{1+x_2})/p_{O_2}(FeO_{1+x_1})]\} \tag{5.27}$$

where $\Delta G(FeO_{1+x})$ is the free energy of the requisite wüstite phase at composition x, and p_{O_2} is the equilibrium partial pressure of oxygen at this composition.

This method has given results of great precision for $\approx$ FeO and for many other non-stoichiometric oxides. The variation of G with composition of the oxides is determined either from cells with successively varied compositions, or by progressively producing small changes in the composition of the oxide in the working electrode by *coulometric titration*. In this procedure, a measured quantity of electricity is passed through the cell by applying an external voltage. When equilibrium is achieved, this corresponds to a measured quantity of oxygen transferred from the standard electrode to the electrode under investigation.

We will include a final example to show that the free energy of formation of complex materials can also be determined using galvanic cells. For this example consider the formation of $CoAl_2O_4$, a spinel which forms from the

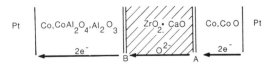

Figure 5.8 Schematic illustration of the processes taking place in a calcia-stabilized zirconia cell used to measure the free energy of formation of $CoAl_2O_4$.

oxides CoO and Al_2O_3 by the following reaction:

$$CoO + Al_2O_3 \rightarrow CoAl_2O_4 \qquad \Delta G_5 \qquad (5.28)$$

The cell to use is shown in Figure 5.8.

The reaction at A is

$$CoO + 2e^- [\leftarrow] \rightarrow Co + O^{2-}[\leftarrow] \qquad (5.29)$$

and at B is

$$CoO + O^{2-}[\leftarrow] + Al_2O_3 \rightarrow CoAl_2O_4 + 2e^- [\leftarrow] \qquad (5.30)$$

Adding (5.29) and (5.30) yields an overall cell reaction identical to that given in equation (5.28). The free energy of formation of $CoAl_2O_4$ from the component oxides CoO and Al_2O_3 will therefore be given by

$$\Delta G_5 = G^0[CoAl_2O_4] = -2EF \qquad (5.31)$$

Many other examples of the use of galvanic cells for free energy measurements will be found by consulting the Supplementary Reading listed at the end of this chapter.

5.4 The β-alumina oxides

5.4.1 *High energy density cells: a problem*

In this section we consider a material which has been very seriously considered for a battery electrolyte in cells primarily intended as power supplies. To do this we first return to equations (5.1) and (5.2) which reveal that in order to obtain a large cell voltage we need an energetic reaction associated with a large free energy change. Typical of such desirable high-energy reactions are those between an alkali metal and a non-metal, for example, Li or Na reacted with a chalcogenide, either O, S, Se and Te, or a halogen, F, Cl or Br. In terms of energy output per unit weight, these reactions yield about 600 kW per hour per kg of material.

The central problem in designing a cell to utilize these energetic reactions is how to separate the highly reactive components of the cell, which become the anode and cathode, by a barrier impermeable to electrons but which can be crossed by ions. One solution to this problem has produced a number of interesting galvanic cells in recent years which rely upon a small family of non-stoichiometric oxides related to the compound β-*alumina* as the electrolyte.

The β-alumina phases are compounds with a general composition lying somewhere between the limits MA_5O_8 and $MA_{11}O_{17}$, where M represents a monovalent cation, typically Li, Na, K, Rb, Ag or Tl, and A represents a trivalent ion, usually Al, Ga or Fe. The parent phase, β-alumina itself, has a nominal composition of $NaAl_{11}O_{17}$.

In fact both the name and formula of β-alumina are misleading. When β-alumina was first prepared it was thought to be a polymorph of alumina, hence the name. It was only later that it was discovered that the compound was, in fact, a ternary sodium aluminium oxide. Despite this fact, the name β-alumina was still retained, and today is firmly entrenched in the literature. As to the composition of the oxide, the compound is a non-stoichiometric phase which contains a variable excess of sodium over and above that given by the proportions $NaAl_{11}O_{17}$.

All the β-alumina-like phases mentioned above are also non-stoichiometric compounds, the variation in composition here being due to a variable alkali metal content. They have high ionic conductivities, with transport numbers for the alkali metal cations close to 1.0 over a wide range of temperatures and oxygen partial pressures as indicated in Figure 5.1 for the sodium compound. It is this high sodium ion conductivity that suggests that the sodium β-alumina phase might be of use as a solid electrolyte in batteries. Before discussing the construction of cells utilizing β-alumina, we will outline some aspects of the structure of this phase which make it so suitable as a solid electrolyte.

5.4.2 The structure of β-alumina

The key to the ability of β-alumina to conduct Na^+ ions so well lies in its structure, and hence to understand its function as a solid electrolyte we must consider this aspect of the material in some detail. The bare bones of the structure were clarified as long ago as 1931. The unit cell is found to be hexagonal, with a c parameter of about 2.25 nm. The structure itself is shown in Figure 5.9. It is seen to be composed of two units, the blocks labelled S and regions between these blocks which hold them together. The S blocks are composed of four oxygen layers in a cubic close-packed arrangement. In these layers, the Al^{3+} ions occupy octahedral and tetrahedral positions, so that the slab has a structure rather like a thin slice of the compound spinel, $MgAl_2O_4$, but without the Mg^{2+} ions. These sheets are often called spinel blocks, but they are neither exactly of the spinel composition, nor blocks, being, in fact, sheets of unlimited extent in the direction normal to the c axis. These rather densely packed slabs are held together only feebly, by two AlO_4 tetrahedra. This means that the sheets are easily separated, and indeed, β-alumina cleaves readily into mica-like foils perpendicular to the c axis. In addition, there is plenty of space between them and it is here that the Na^+ ions reside.

When the structure is determined carefully, problems arise. For electrical

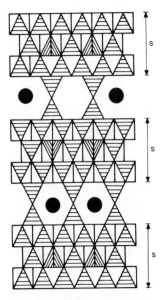

Figure 5.9 The structure of β-alumina shown as a packing of spinel layers, labelled S, separated by open sheets in which the alkali metal ions, shown as circles, are to be found.

neutrality an ideal formula of $Na_2O.11Al_2O_3$ is needed, giving a unit cell content of $Na_2Al_{22}O_{34}$. If we try to locate the sodium atoms precisely it proves to be impossible, and it appears that we have a random distribution of Na^+ ions over the available sites which exist in the layers between the spinel slabs. As the layer containing the sodium atoms is by no means filled, the Na ions can be thought of as moving continuously from one position to another, rather like the situation that we met in Chapter 1 with α-AgI. This means that the Na^+ ions are free to move in the plane and we can consider the Na^+ ions to behave as a quasi-liquid layer. The structure of this Na^+-containing plane is shown in Figure 5.10. It is now easy to understand why β-alumina is such a

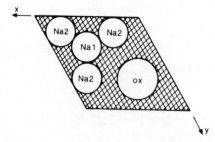

Figure 5.10 The possible Na sites in sodium β-alumina. The positions labelled Na1 have a fractional occupancy of about 0.8, while the sites labelled Na2 have a fractional occupancy of about 0.166. The circle labelled ox shows the position of normal oxygen ions in the layer. The shaded outline marks the extent of the unit cell.

good conductor of Na^+ ions. We have almost unimpeded motion in the Na^+ layers and the conductivity is of the same order of magnitude as one would find in a strong solution of a sodium salt in water. However, we can see that conduction takes place only in a plane normal to the c axis.

The complex story of the non-stoichiometric structure of β-alumina is not quite finished, however. While it has an ideal composition of $NaAl_{11}O_{17}$, the real composition is quite variable and shows that the phase always contains an excess of alkali metal. A typical analysis would yield a composition of $Na_{2.58}Al_{21.81}O_{34}$, for example. Because crystals of β-alumina contain an excess of Na^+ ions over the idealized formula, it is necessary to look for some sort of counter defects. There are two reasonable possibilities that can be envisaged: the introduction of Al^{3+} vacancies into the spinel sheets or else the incorporation of extra oxygen ions into the structure. These extra O^{2-} ions cannot be introduced into the spinel sheets as they are already made up of fully close-packed layers of oxygen ions, which means that the oxygen can reside only in the same layers as the Na^+ ions. The resolution of this problem is not easy and the results are not altogether unequivocal. However, it seems that in β-alumina it is this latter solution of the problem which holds, and some interstitial oxygen ions enter the Na planes to maintain the charge balance. In some of the other related phases, cation vacancies or a combination of both vacancies and interstitial oxygen may occur.

We can see, therefore, that in β-alumina, the problem of high point defect concentrations has been solved by effectively segregating the 'defects' into Na^+-containing layers, away from the normal spinel-like structure which contains only low point defect concentrations. This is indeed a clever structural way for aluminium oxide, which is built up of small cations, to incorporate large cations within its structure without excessive lattice strain. Hence the phase has resorted to a quite different strategy to that found in calcia-stabilized zirconia to overcome the problem of accommodating large defect populations.

We can also note that the ionic conductivity will be specific to the alkali metal cations between the spinel slabs. Thus, if we wish to make a sensor capable of detecting only sodium we could make use of sodium β-alumina. If we substitute the Na-containing phase for one of the isostructural phases containing different alkali metal atoms, then we change the specificity of the sensor.

5.4.3 β-alumina secondary storage cells: a solution

As the ionic conductivity of Na^{1+} ions in β-alumina is close to that of typical liquid electrolytes, while the phase shows no significant degree of electronic conductivity under normal circumstances, it seems possible that this material might be suitable to separate the reactants in the high energy density cell that was mooted in section 5.4.1. Cells in which the reaction is between Na and a chalcogen could therefore use β-alumina as a solid electrolyte. A cell of this

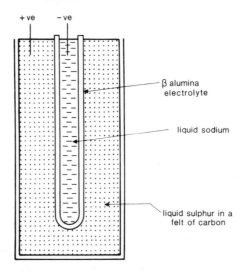

Figure 5.11 Schematic diagram of a sodium–sulphur cell using β-alumina as a solid electrolyte.

type was first demonstrated by the Ford Motor Company in Detroit in 1966. As expected, it had an extremely high power density, equal to $1030\ \mathrm{Wh\,kg^{-1}}$.

The principle of the cell is shown in Figure 5.11. The reaction chosen was that between sodium and sulphur. The β-alumina electrolyte separates molten sodium from molten sulphur, which is contained in a porous carbon felt. The operating temperature of the cell is high, about $300\ ^\circ$C, which is something of a drawback. However, the cell reaction is extremely energetic, and the heat required to maintain the cell at its operating temperature is readily supplied by the cell reaction itself.

In the cell, the following reactions take place. At the liquid sodium–β-alumina interface sodium atoms lose an electron and enter the Na$^+$ layers in the crystal, thus:

$$2\mathrm{Na} \rightarrow 2\mathrm{Na}^+ + 2e^- \qquad (5.32)$$

At the interface between the β-alumina and the liquid sulphur we have a complex reaction to form polysulphides of sodium. As an example of this type of reaction we will write

$$2\mathrm{Na}^+ + \mathrm{S} + 2e^- \rightarrow \mathrm{Na_2S} \qquad (5.33)$$

The overall cell reaction

$$2\mathrm{Na} + \mathrm{S} \rightarrow \mathrm{Na_2S} \qquad (5.34)$$

will take place as Na$^+$ ions move freely across the β-alumina via the conducting planes.

Since 1966, much effort has gone into producing commercial batteries based on the sodium–sulphur cell. In this enterprise small differences in the electrical properties of the electrolyte can be of major significance. Because of this a number of other β-alumina-like structures have been made in which the stacking of the spinel layers is slightly different than in sodium β-alumina, or else the thickness of the spinel layers increases slightly. Of these other materials, one in particular, referred to as β''', is of great importance in applications. Nevertheless the principle of operation of the cells with these other electrolytes remains precisely the same as that described above.

5.5 Lithium nitride cells

The compound lithium nitride has been of potential interest as a solid electrolyte for a number of years, but it is not this aspect of the material that we will concentrate on in this section. However, by way of an introduction it is of interest to consider how the material functions as an electrolyte. For this it is necessary, as usual, to look at the structure. It is found that the phase is composed of compact layers of Li and N atoms joined together by Li atoms between the layers. There is some uncertainty about the bonding in Li_3N, but is would appear that an ionic model is not too far from the truth. In this case, the layers have a formula of $(Li_2N)^-$ and they are linked by Li^+ ions. The structure is illustrated in Figure 5.12.

Pure Li_3N is a conductor of Li^+ ions. Looking at Figure 5.12 it would be natural to assume that the ionic transport takes place exactly as in β-alumina and that the inter-layer lithium ions are responsible for the conductivity. This, however, is false, as the phase is strictly stoichiometric. Instead, the conductivity comes about by the formation of Frenkel defects involving the Li^+ ions in the hexagonal layers. Some of these move into the region between the layers as interstitials, leaving a vacancy behind them. The energy required to form such a defect is only 0.19×10^{-19} J. Surprisingly, the migrating defects appear to be the Li vacancies and not the interstitial Li ions between the layers. The enthalpy of migration of the defects is about 0.19×10^{-19} J, the same as the formation energy of the Frenkel pair.

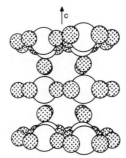

Figure 5.12 The structure of Li_3N. The hexagonal structure contains layers of Li^+ ions, shown as small spheres, and N^{3-} ions, shown as large spheres, connected by bridging Li^+ ions.

In order to make a useful battery the Li_3N is employed as the electrolyte to separate the reactive components of the cell. As the electrolyte in this case is a Li^+ ion conductor, Li will be one component. The other component will be anything which is able to react with lithium to produce a stable phase. The considerations set out earlier suggest that this will be an element from the right of the Periodic Table, and sulphur would appear to be a reasonable choice. In practice, sulphur does not react very rapidly when cold, and although this is not a problem in a cell which is to be used for motive power, such as the sodium β-alumina battery described above, it is an important constraint in small cells for use in electronics applications. The ideal electrode materials for this purpose have been found to be some non-stoichiometric disulphides of transition metals, such as TiS_2, the structure of which is shown in Figure 4.4. The reaction with Li^+ ions results in the incorporation of variable amounts of Li atoms between the TiS_2 layers to form non-stoichiometric Li_xTiS_2.

The cell reactions would be as follows.
At the anode:

$$xLi \rightarrow xLi^+ + xe^- \qquad (5.35)$$

At the cathode:

$$TiS_2 + xLi + xe^- \rightarrow Li_xTiS_2 \qquad (5.36)$$

giving the overall cell reaction

$$xLi + TiS_2 \rightarrow Li_xTiS_2 \qquad (5.37)$$

which takes place for values of x which are less than one. The Li^+ ions are transported via the Li_3N, while the electrons traverse the external circuit. The non-stoichiometric nature of the cathode material is of importance because any amount of Li transferred by the electrolyte can be incorporated into the sulphide, which can be regarded as a variable reservoir of Li. These cells present the rather unusual feature, therefore, that a non-stoichiometric compound is used for the cathode and a stoichiometric material for the electrolyte. A diagram of such a cell is shown in Figure 5.13.

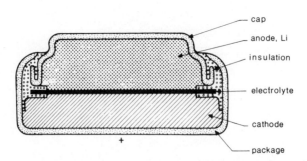

Figure 5.13 Schematic illustration of a Li_3N solid state battery.

5.6 Supplementary reading

There are a number of textbooks, volumes of conference proceedings and articles devoted to solid state electrolytes which cover the topic in this chapter from the point of view of cells and batteries. A selection of these are: *Solid Electrolytes* (Topics in Applied Physics), ed. S. Geller, Springer Verlag (1977); *Solid Electrolytes: General Principles, Characterisation, Materials, Applications*, eds. P. Hagenmuller and W. van Gool, Academic Press, London (1978); *Fast Ion and Mixed Conductors*, Risø International Symposium on Metallurgy and Materials Science, eds. F.W. Poulsen, N.H. Andersen, K. Clausen, S. Skaarup, and O.T. Sørensen, Risø National Laboratory, Denmark (1985); 'Solid state electrochemistry', H. Rickert, in *Treatise on Solid State Chemistry*, Vol. 4, ed. N.B. Hanny, Plenum, New York (1976); *Modern Batteries*, C.A. Vincent, Edward Arnold, London (1984).

There are a number of shorter, very readable articles available, such as M.D. Ingram and C.A. Vincent, 'Solid state ionics', *Chem. in Britain* **20** (1984) 235; A. Rabenau, 'Lithium nitride and related materials', *J. Ed. Mod. Mater. Sci. Eng.* **4** (1982) 493; S. Skaarup, 'Solid electrolytes', *J. Mater. Ed.* **6** (1984) 667.

These last two are highly recommended as lucid and informative articles.

Part 3 Non-stoichiometric materials containing atoms of variable valence

6 Non-stoichiometry and electronic conduction in materials containing ions of variable valence

6.1 Introduction

So far in this book we have restricted discussion to materials in which the cations took only one valence state. We now add another level to our understanding of the structure and properties of non-stoichiometric compounds by considering materials in which some of the cations can take more than one valence state. Such cations are typically those of the transition metals.

Naturally, all of the features of point defect chemistry that have been discussed previously also pertain to transition metal compounds. Thus, non-stoichiometric phases can contain substituted atoms, vacancies or interstitials, and we know that non-stoichiometry of this sort must also be accompanied by compensating defects which maintain overall electrical neutrality. Previously these compensating defects were also supposed to be vacancies, interstitials or substituted atoms. However, we now have another possibility, that electrical neutrality can be preserved internally, simply by redistributing the valences of the cations in the materials. This is the key to understanding the important electrical properties of the transition metal compounds and the way in which these properties can be manipulated for our own purposes. In this and subsequent chapters we will discuss these properties and some of the more important features of non-stoichiometric phases that are derived from them.

6.2 Non-stoichiometry in pure materials

We will consider here how the fact that some ions possess a variable valence can result in the formation of new sorts of defects in crystals of transition metal compounds. These defects are electronic in nature and are created in response to the problem of keeping the crystals neutral over a range of compositions. In order to keep this clearly in mind, we will ignore the presence of Schottky and Frenkel defects, which will not alter the stoichiometry of the crystals, and will also not consider compensation by vacancies or interstitials. These extra complexities cannot, of course, be ignored in real crystals. In addition, we will essentially treat systems which show only small deviations from stoichiometry.

119

As an example, let us look at the way in which an oxide of ideal formula MO can accommodate variation in composition. Consider first an oxide with an experimentally determined metal excess compared to the anion content, that is, the cation: anion ratio is greater than one. We can allow for this in two ways.

(i) Type A materials. If anion vacancies are present the oxide will have a real formula MO_{1-x}.
(ii) Type B materials. If interstitial cations cause the excess metal, the oxide will have a real formula of $M_{1+x}O$.

We can illustrate these models schematically in Figure 6.1, although, as we have stressed before, it must be remembered that such pictorial representations are certain to be far from the truth in a structural sense.

The most important feature to note is that, in the absence of other vacancies or interstitials, we cannot make these compositional changes without introducing *electronic* defects as well. For example, in type A materials with anion vacancies, if we simply remove anions the crystals will end up with an overall positive charge. In order to keep the crystal neutral we need to introduce two electrons for each oxygen ion moved. As such an anion vacancy has an effective positive charge, we have shown the two electrons to be trapped at the anion vacancy in Figure 6.1(a). In practice this may or may not happen, and the actual location of the electrons will depend upon the bonding in the crystal and the geometry of the vacant sites.

The introduction of electrons into a crystal is often, however, an extremely energetic process, and it does not always occur readily. An alternative is for the electrons to be associated with a cation. In our case if one electron is associated

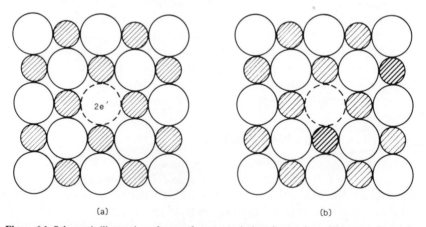

(a)   (b)

Figure 6.1 Schematic illustration of ways of accommodating changes in stoichiometry in a cation excess oxide of composition close to MO by way of anion vacancies. The normal anions and cations are represented as full circles and the vacancy by a dotted circle. Two electrons have been introduced into the crystal to maintain charge neutrality. These can be localized at the vacancy, as in (a), or distributed in other ways within the crystal. In (b) the electrons are supposed to be sited on two normal cations making them into M^{1+} ions, which are shown heavily shaded.

with one cation, it will change from an M^{2+} ion to an M^{1+} ion. This process will occur only if the relevant ionization energies are small. In practice this means that it will only occur for ions which have a variable valence, i.e. transition metal ions. Thus we find that because of the ability of a transition metal to change its valence readily, a pure transition metal compound has the capability to accommodate stoichiometric variation with only a small energy expenditure. This situation is illustrated in Figure 6.1(b).

A similar situation can hold with the second group of materials, type B. In order to maintain charge neutrality we need to introduce interstitial atoms which are neutral entities. Now it is reasonable to suppose that interstitial atoms as such are not likely to exist in an oxide of composition close to MO, and they are likely to lose electrons to form interstitial ions. In the present case each interstitial M atom must donate two electrons into the crystal. These electrons can remain associated with the interstitial cation, as shown in Figure 6.2(a), or else they could be sited on normal lattice cations to create a random or ordered array of M^{2+} ions in the structures, as shown in Figure 6.2(b).

In these materials non-stoichiometry involves the introduction of electrons which are not necessarily strongly bound to particular atoms into the crystal. They can, therefore, conduct electricity if energy is supplied to free them from the places where they may be trapped. Thermal energy is often sufficient for this purpose, and in such cases the electronic conductivity will rise with increasing temperature. This behaviour is typical of semi-conductors, and as

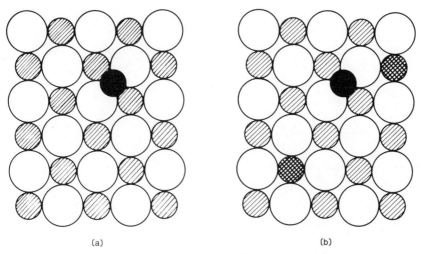

(a)   (b)

Figure 6.2 Schematic illustration of ways of accommodating changes in stoichiometry in cation excess oxides of composition close to MO by way of interstitials. The normal anions and cations are represented by circles. In (a) the interstitial atom, which still has its valence electrons associated with it, is shown black. In (b) the atom has ionized to M^{2+}, shown black, and two electrons have been liberated into the crystal. These have been located on two normal M^{2+} cations to create two M^{1+} ions, shown cross-hatched.

E

the liberated charge carriers will be the trapped electrons, the materials will be *n*-type. The energy to liberate the electrons can also, in certain cases, be provided by photons, which results in photoconductivity and sometimes causes the materials to become coloured. This aspect is discussed later in this volume.

Not all non-stoichiometric oxides are cation-rich, of course. If we have an oxide in which we find that we have more anions than cations, i.e. in which the cation: anion ratio is less than one, we have two analogous methods of accommodating this structurally.

(i) Type *C* materials. Interstitial anions will cause non-metal excess, giving the oxide a real formula of MO_{1+x}.
(ii) Type *D* materials. Alternatively, cation vacancies can result in non-metal excess, giving the oxide a formula of $M_{1-x}O$.

These are shown schematically in Figure 6.3.

In the cases of these type *C* and *D* materials the situation is somewhat similar to that already described for the *A*- and *B*-type compounds. In the present case, though, electrons must be removed to maintain charge neutrality. In type *C* materials, for example, each additional O^{2-} interstitial introduced necessitates the removal of two electrons to keep the crystal neutral. These are most readily supplied by a cation, and so Figure 6.3(*a*) shows that two M^{2+} cations have been converted to two M^{3+} cations which have been drawn close to the interstitial oxygen. This process will occur only if the cations are able to

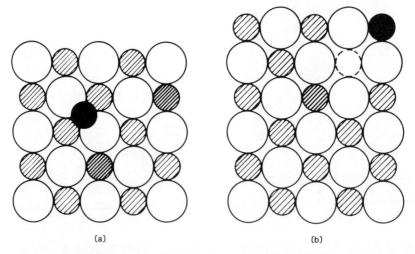

(a)   (b)

Figure 6.3 Schematic illustration of structural ways of accommodating changes in stoichiometry in anion excess oxides: (*a*) by way of interstitial anions, shown as filled circles, plus altervalent M^{3+} cations, shown shaded; (*b*), by way of cation vacancies plus altervalent M^{3+} cations, shown shaded. The altervalent cations may be either distributed at random in the structure or else localized near the defect which creates them.

gain or lose electrons easily, and so once again we find that this sort of non-stoichiometric behaviour is restricted to compounds of the transition metals.

The same can be said of the situation shown in Figure 6.3(b) where we have cation vacancies. Each missing cation needs to be balanced by the introduction of two extra positive charges, which change two M^{2+} ions into two M^{3+} ions, as shown. It goes without saying that the position of the M^{3+} ion will depend on a variety of factors, and they may be situated close to the defect which produces them, or in quite a different part of the crystal.

Just as we regarded an M^{2+} ion plus an electron as an M^{1+} ion, we can regard the M^{3+} ions in type C and D materials as M^{2+} ions plus a trapped positive hole. As before, thermal energy may be able to liberate these positive holes. This will result in electronic conductivity so that these materials will fall into the group of p-type semiconductors. The charge transfer in p-type materials is still, of course, by way of electrons, and regarding the mobile charge carriers as positive holes is for convenience.

Examples of materials which can be placed into these classification are few, as the structures of many non-stoichiometric compounds are more complex than that adopted in our present simple description. However, type B, n-type semiconducting materials are well exemplified by ZnO and CdO, while type D, p-type semiconducting materials are exemplified by the oxides NiO, Cu_2O, CoO and MnO.

Clearly in all these materials the number of electronic defects is closely related to their composition. Thus, by controlling stoichiometry, we can control electronic conductivity, a fact of some importance in the potential production of materials of use in electronic devices. For example, NiO forms the basic material employed in many thermistors; NiO and MnO together with FeO are important components of ferrites; while the non-stoichiometric behaviour of UO_2 is clearly of considerable interest in nuclear fuel technology.

6.3 The effect of impurity atoms

Although in principle it is not difficult to envisage how the stoichiometry of a pure phase can be changed to enhance its electronic properties, following the outlines given above, in practice it is less straightforward. As discussion later in this book will make clear, the composition of a pure non-stoichiometric compound depends upon the partial pressure of the vapours surrounding the crystal. Thus, in order to control stoichiometry we also need to control the atmosphere in contact with the phase. This is impractical for many purposes. Moreover, the existence range of any particular non-stoichiometric material is often very limited. The composition range available for a phase such as NiO, for instance, may simply be too narrow to encompass the range of properties which we wish to exploit.

The importance of these materials makes it worthwhile searching for simpler ways of obtaining the desired electrical properties without the constraints of always having to work in a controlled partial pressure of non-metal and metal vapour. Fortunately the same electronic effects are generated if we can incorporate into the lattice an impurity ion of a different nominal charge or valence to that of the parent atoms. This also frequently allows one to extend the non-stoichiometric behaviour of the phase considerably, since altering the partial pressure of oxygen or metal vapour over an oxide generally leads to only small departures from stoichiometry, whereas several per cent of impurity ions can often be accommodated in a crystal.

Thus, heating Li_2O and NiO together at high temperatures will cause the Li^{1+} ions to enter the NiO structure. This is able to take place relatively easily because the ionic radii of Ni^{2+} and Li^{1+} in the octahedral coordination found in NiO are very similar. The resulting material is black and can be described by the formula $Li_xNi_{1-x}O$, where x can take values from 0 to about 0.1. As the Li^{1+} ions are introduced into the NiO crystals a compensating defect is needed to balance the charge and maintain neutrality. This turns out to be Ni^{3+}. Thus every one Li^{1+} in the lattice produces one Ni^{3+} ion. As we noted in the previous section, this is equivalent to a high concentration of holes located on Ni^{2+} cations. The process of creating electronic defects in a crystal in this way is called *valence induction*.

In the process of valence induction we are simply duplicating the effect of the normal variation in composition found in NiO by using another technique, but the effective composition range in this synthetic compound is far greater than that in pure NiO itself. As expected, the electronic conductivity is appreciable and can be made to approach that of a metal at higher Li concentrations. Nevertheless the conductivity still increases with temperature, indicative of a semiconductor rather than a true metal. It should be noted, of course, that if a change of valence for at least one of the cations was not possible, we would have to make use of vacancies or interstitials to maintain the charge neutrality, as described in Chapter 4. This would not produce electronically conducting materials, of course, but insulators.

It is equally possible to enhance *n*-type conductivity by suitable doping. For example, consider the consequences of reacting Ga_2O_3 with ZnO. The Ga ions have a charge of 3 +, and hence for each Ga^{3+} which enters the ZnO structure some charge compensation is needed. If the Ga^{3+} substitutes for Zn^{2+} and the crystal maintains its overall MO stoichiometry, we will lose oxygen during the reaction. The electrons so made available will then be able to remain in the structure to affect the necessary charge compensation. We can write this in the form of a chemical equation, thus:

$$Ga_2O_3 \xrightarrow{ZnO} 2Ga_{Zn}^{\cdot} + 2O_O + \tfrac{1}{2}O_2(g) + 2e'$$ (6.1)

Each Ga^{3+} in the lattice is balanced by an electron somewhere which may rest on a Zn^{2+} to generate either Zn atoms or maybe Zn^{1+} ions. We can write this as

$$Zn_{Zn} + e' \rightarrow Zn'_{Zn} \tag{6.2}$$

$$Zn_{Zn} + 2e' \rightarrow Zn''_{Zn} \tag{6.3}$$

We will encounter the effects of impurities often in the remainder of this book and hence further details will not be given in this brief section. However, it is important to be aware that deliberate doping with impurities to cause specific changes in electronic properties of materials lies at the heart of much of the modern electronics industry.

6.4 Electronic conduction in materials with localized charge carriers

The electronic properties of materials are determined by the behaviour of the mobile electrons in the solid. For a large class of materials, especially the metals, band theory is pre-eminent in explaining these properties. In band theory the electrons responsible for conduction are assumed to be free or very nearly so. To put this another way, the wave functions of these electrons are considered to be delocalized and to extend throughout the whole of the crystal. Band theory is well explained in many textbooks of solid state physics and it would be unnecessary to repeat the fundamentals here. For those who need to know more about the band theory equations mentioned in this chapter, sources are given in section 6.7, Supplementary Reading.

Now the materials described in the earlier sections of this chapter have been discussed in terms of an ionic model. It would therefore be useful to set up a theory for electronic conductivity that retained this simple picture. In the remainder of this chapter we describe such an alternative theory and show how it is able to account for some of the interesting electronic properties of non-stoichiometric compounds that are not readily obtained via band theory. Despite this success it would be artificial to ignore band theory altogether, and in the final section of this chapter the two approaches are linked together, at least qualitatively.

The ionic model of solids treats the electrons in the material as being localized at ions and not free to move at all. In this section the conduction electrons are considered to be additional electrons or holes created by the defect chemistry of the system, and mostly localized or trapped at atoms or other defects within the crystal. They contribute to electronic conductivity by jumping or *hopping* from one site to another under the influence of an electric field. At a certain time, a localized electron will acquire enough energy to overcome the trapping barrier. It will then move to another site where it becomes relocalized until it gains sufficient energy to make another jump.

If this is so, one would expect that oxides with only one valence state, such as stoichiometric MgO, would be insulators. This is because if we want to move an electron from one cation to another we must provide energy equal to the further ionization energy of one Mg^{2+} cation, say, and the electron affinity of another Mg^{2+} cation, to result in the hypothetical production of Mg^{3+} and Mg^{1+}, as shown in Figure 6.4(a). Such a situation requires so much energy that it is never encountered under normal circumstances, and the material will remain an insulator. The same applies, of course, to the anions. On the other hand, if two valence states are normally available, as in non-stoichiometric

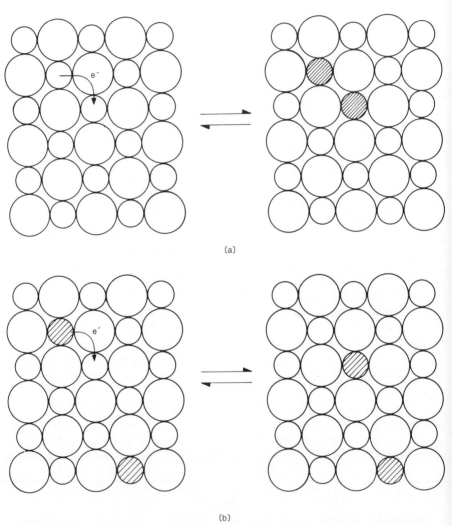

Figure 6.4 In (a), representing a stoichiometric oxide, with cations of fixed valence, say Mg^{2+}, the electron jump shown arrowed requires a large energy input and creates an Mg^{1+} and an Mg^{3+} ion, shown shaded. For a slightly non-stoichiometric oxide, illustrated in (b), already containing a low population of M^{3+} ions, shown shaded, the electron exchange is relatively easy, as the initial and final states of the crystal are very similar.

$Ni_{1-x}O$, which contains Ni^{2+} and Ni^{3+} ions, an electron jump requires very little energy, because the initial and final state of the crystal are very similar, as shown in Figure 6.4(b). Thus under normal conditions, electron movement will not be too difficult, and the material will be an electronic conductor of some sort. Electronic conductivity by a hopping mechanism is therefore likely to be restricted mainly to transition metal compounds where alternative valence states are available to cations with little expenditure of energy.

A little thought will show that this electron movement by way of discrete jumps is identical to that of atom diffusion discussed in Chapter 3. Thus, in contrast to band theory, conduction in materials with hopping charge carriers is essentially a diffusion process. As treatment of diffusion in this earlier chapter was successful in accounting for many aspects of atom movement, it is worthwhile applying it to the present problem to see exactly where it will lead us.

Referring to Chapter 3, we know that for a random diffusion process the relationship between ionic conductivity, σ, and self-diffusion coefficient, D, is given by the Einstein relation, viz.

$$\sigma = ne^2 D/kT \qquad (6.4)$$

where n is the number of mobile charge carriers per unit volume. As the hopping process will be identical to the diffusion process, equation (6.4) will also apply to electron or hole hopping, provided that it takes place by a random series of electron jumps.

The strategy that is now employed is to derive a theoretical expression for the diffusion coefficient, following the route laid down in Chapter 3, and substitute this in equation (6.4) to obtain a relationship between the conductivity, σ, and other atomic parameters. We therefore start with the expression for the number of successful jumps made by a hopping electron, which is

$$q = v\exp[-E/kT] \qquad (6.5)$$

where v is the attempt frequency for a hop, q is the probability that a jump along the field direction will be successful, and E is the activation energy for the hop.

Now such an expression is valid if each possible jump is to an available site. In the case under discussion the electron cannot just jump anywhere. If we consider our example of $Ni_{1-x}O$, an electron can jump from Ni^{2+} to Ni^{3+} but not from one Ni^{2+} to another or from one Ni^{3+} to another. To allow for this, we can designate the number of sites which are occupied by mobile charges, either e' or $h^{\cdot}$, by ϕ, which is expressed as a fraction of the total sites which the mobile charge carriers are allowed to occupy. Thus, the fraction of available unoccupied sites is $(1 - \phi)$. With this proviso we can now rewrite equation (6.5) in the correct form for us, so that the probability of a successful jump will be given by

$$q = (1 - \phi)v\exp[-E/kT] \qquad (6.6)$$

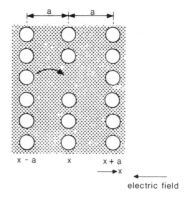

Figure 6.5 Schematic illustration of electron 'diffusion' between three adjacent crystal planes, located at $(x - a)$, x and $(x + a)$. Because the electrons carry negative charges they will diffuse in a direction opposite to the electric field direction.

We now calculate the diffusion coefficient of these moving charges, following the procedure in Chapter 3. The steps in the argument will therefore be but briefly outlined. In the electric field responsible for the electronic conductivity a gradient of mobile charge carriers will exist. This is indicated schematically in Figure 6.5, which shows three adjacent lattice planes in the crystal under consideration, separated by a jump distance a. If we have a density of na carriers on the plane at position x, at time t, per unit volume, and a density of $[n + a(\mathrm{d}n/\mathrm{d}x)]a$ and $[n - a(\mathrm{d}n/\mathrm{d}x)]a$ on the two adjacent planes at time t, we can write:

$$\mathrm{d}n/\mathrm{d}t = \tfrac{1}{2}q(1 - \phi)a^2\mathrm{d}n/\mathrm{d}x \tag{6.7}$$

Now Fick's law in the form that we require is

$$\mathrm{d}n/\mathrm{d}t = D\,\mathrm{d}n/\mathrm{d}x \tag{6.8}$$

so that we can write

$$D = \tfrac{1}{2}q(1 - \phi)a^2 \tag{6.9}$$

where D is the diffusion coefficient of the charge carriers. Substituting for q from equation (6.6) allows us to write

$$D = \tfrac{1}{2}v(1 - \phi)a^2 \exp[-E/kT] \tag{6.10}$$

We are now able to go back and substitute this into the Einstein relation, equation (6.4), to yield

$$\sigma = \{ne^2v(1 - \phi)a^2 \exp[-E/kT]\}/2kT \tag{6.11}$$

This expression is rather cumbersome and it is worth our while to condense it somewhat. To do this let us look at the basic building blocks of the structure through which the electrons are diffusing, that is, the unit cells. If we have c

sites that the mobile charge carriers can occupy per unit cell, of volume v, then the number of mobile charge carriers per unit cell will be $c\phi$, and the number per unit volume will be

$$n = c\phi m/v \qquad (6.12)$$

where m is the number of unit cells per unit volume. If we now take the vibration frequency to be independent of temperature, we can collect many of these terms into a constant factor K, and write equation (6.11) in the form

$$\sigma = K(1 - \phi)\exp(-E/kT) \qquad (6.13)$$

where

$$K = cma^2ve^2/2vk \qquad (6.14)$$

In this equation, ϕ is the equation of sites occupied by a mobile electron, $(1 - \phi)$ is the fraction of unoccupied positions that the electron can move to and E is the activation energy for each jump at a temperature T.

We see that the conductivity, σ, is an activated process, that is

$$\sigma \propto \exp - E/kT \qquad (6.15)$$

This means that the conductivity will increase with temperature, and so hopping materials will be semiconductors. Hopping materials are thus often referred to as *hopping semiconductors*.

However, the conductivity will also vary as a function of ϕ, and this is something new. To illustrate the implication of this fact, consider a non-stoichiometric oxide AO_x in which x can take all values between one and two. Within this composition range, suppose that three stoichiometric oxides AO, A_2O_3, and AO_2 form. These contain, nominally, A^{2+}, A^{3+} and A^{4+} cations. How exactly will the electronic conductivity vary over the total composition range of the non-stoichiometric phase?

To arrive at an answer to this question, suppose that stoichiometric AO_2 is heated in a vacuum so that it loses oxygen. Initially, all cations are in the A^{4+} state and we expect the material to be an insulator. Removal of O^{2-} to the gas phase as oxygen causes electrons to be left in the lattice, which will be localized on cation sites to produce some A^{3+} cations. The oxide now has a few A^{3+} cations in the A^{4+} matrix, and thermal energy should allow electrons to hop from A^{3+} to A^{4+}. Thus the oxide should be an n-type semiconductor. The conductivity becomes easier until $x = 1.75$, when there are equal numbers of A^{3+} and A^{4+} cations present. As reduction continues, we get to a stage when almost all the ions are now in the A^{3+} state and we have only a few A^{4+} cations left. This is analogous to the situation in $Ni_{i-x}O$, and the material should be a p-type semiconductor. Eventually at $x = 1.5$, all cations will be in the A^{3+} state and we have an insulator, A_2O_3.

We can repeat this argument for the composition range from A_2O_3 to AO. Slight reduction of A_2O_3 will produce a few A^{2+} cations in the A^{3+} matrix, leading to n-type semiconductivity. This would persist in the composition

conductivity
σ

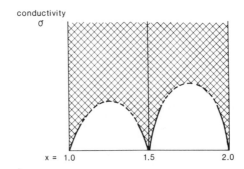

x = 1.0   1.5   2.0

Figure 6.6 Diagrammatic illustration of the change of conductivity with composition expected for hopping semiconductor.

range AO_x between $x = 1.5$ and $x = 1.25$, the conductivity passing through a maximum at the composition $AO_{1.25}$. Further reduction would lead to the situation where we have fewer A^{3+} cations than A^{2+}, and we expect p-type behaviour in the composition range between $x = 1.25$ and $x = 1.0$. The stoichiometric composition AO should be an insulator.

The simple arguments proposed here suggest that conductivity of the material should behave in the fashion shown schematically in Figure 6.6. There are no non-stoichiometric phases with such a wide composition range, but many which cover parts of it. These show that the conductivity of such phases does vary in the way expected. This gives credence to the idea that hopping conductivity does occur in some phases, and that a diffusion model offers a reasonable explanation of the experimental findings.

In addition to a knowledge of conductivity, especially for semiconductor materials, we need information about the mobility of the charge carriers. It is useful, therefore, to also briefly consider this aspect for our hopping model. We can recall that the basic equation which relates the conductivity of a material, σ, to the number of mobile charge carriers in the material, n and their mobility μ, is

$$\sigma = ne\mu \qquad (6.16)$$

where e is the charge on the mobile species, which in this case is simply the charge on an electron. We can then go back to equation (6.13) and write, for a hopping material,

$$\mu = [K\phi(1 - \phi)/neT]\exp(-E/kT) \qquad (6.17)$$

Hence

$$\mu \propto \exp(-E/kT) \qquad (6.18)$$

and the mobility is an activated process. It increases as the temperature goes up, and so for a hopping semiconductor σ increases with T because μ increases with T. The number of mobile carriers on the other hand, depends on the composition of the crystal, as was illustrated in Figure 6.6. If the composition is fixed then n does not vary with temperature, but the conductivity still does via the mobility.

We can note that the type of behaviour predicted by equations (6.13) and (6.17) is quite different from that of a band-like material, such as an alloy of two metals. With such a material, for example, the conductivity is frequently a linear function of the composition and certainly does not tend to zero at the end compositions. Thus, measurements of conductivity and mobility are able to provide quite useful information about the non-stoichiometric nature of the material under investigation. Because of this it is useful to compare the behaviour expected in terms of band theory with that predicted by our simple equations (6.13) and (6.17).

For a typical band-theory metal, the conductivity is given by

$$\sigma = ne^2\tau/m^* \qquad (6.19)$$

and the mobility by

$$\mu = e\tau/m^* \qquad (6.20)$$

In these equations, n, the number of charge carriers in the metal, is constant, m^* is the effective mass of the electron in the metal and τ is the mean lifetime between electron collisions. At normal temperatures μ is proportional to a low power of temperature:

$$\mu \propto T^{-m} \qquad (6.21)$$

where m is typically about 3/2. For a metal, therefore, σ decreases with temperature, as

$$\sigma \propto ne/T^m \qquad (6.22)$$

In the case of semiconductors which are best treated in terms of band theory the same transport equations apply, i.e.

$$\sigma = ne^2\tau/T^m \qquad (6.23)$$

and

$$\mu = e\tau/m^* \qquad (6.24)$$

The mobility is, therefore, dependent upon temperature as in a metal, i.e.

$$\mu \propto T^{-m} \qquad (6.25)$$

In these materials, the conductivity rises with temperature because n is sensitive to temperature. This dependence is well known and is usually written as

$$n = n_0 \exp\left[E_F - E_g\right]/kT \qquad (6.26)$$

for electrons, and

$$n = n_0 \exp\left[-E_F/kT\right] \qquad (6.27)$$

for holes, where E_F is the Fermi energy of the semiconductor, E_g is the band-

Table 6.1 Electrical conductivity and mobility of charge carriers in metals, band-like semiconductors and hopping semiconductors

	Metal	Band-like semiconductor	Hopping semiconductor
Conductivity, σ	$\propto T^{-m}$ Falls slightly with T	$\propto \exp[-E/kT]$ Increases with T	$\propto \exp[-E/kT]$ Increases with T
Mobility, μ	$\propto T^{-m}$ Falls slightly with T	$\propto T^{-m}$ Falls slightly with T	$\propto \exp[-E/kT]$ Increases with T

gap of the material and n_0 is a constant. Thus, we find that the conductivity increases with temperature in a semiconductor, because the value of n increases in the fashion described by equations (6.23) and (6.24), so that

$$\sigma \propto \exp(-E/kT) \tag{6.28}$$

and

$$\mu \propto T^{-m} \tag{6.29}$$

A comparison of the relevant equations for hoppers, for metals and for a band-theory semiconductor therefore shows that if the conductivity increases with temperature we either have a band-like semiconductor or a hopper, but not a metal, and if the mobility increases with temperature we have a hopping conductor. These distinctions are summarized in Table 6.1.

6.5 Thermoelectric effects: the Seebeck coefficient

Temperature gradients in materials with mobile charge carriers usually lead to gradients in electrical potential. These effects are termed *thermoelectric effects* and they can give information about the nature and the number of charge carriers present in a material. In this section we discuss one thermoelectric effect, the *Seebeck effect*, and show how it is able to yield such information.

When the two ends of a conductor or semiconductor are held at different temperatures, as shown schematically in Figure 6.7, a potential develops across the material. It is this phenomenon which is called the Seebeck effect.

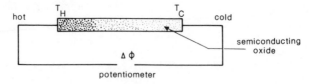

Figure 6.7 Schematic demonstration of the Seebeck effect. The sample, which is typically an oxide such as NiO, is placed in a temperature gradient so that the temperature varies from one end, which is at T_H, to the other at T_C. This results in a potential difference of $\Delta\phi$ between the ends when equilibrium is reached.

Besides being the physical principle which underlines the use of thermocouples for temperature measurements, the effect is widely used to characterize the charge carrier type in oxide semiconductors. It is found that the potential developed, $\Delta\Phi$, is given by

$$\Delta\Phi = \Phi_H - \Phi_C = \pm\, \alpha(T_H - T_C) = \pm\, \alpha\Delta T \qquad (6.30)$$

where Φ_H and Φ_C are the potentials and T_H and T_C are the temperatures at the hot end and the cold end of the sample respectively. The Seebeck coefficient, α, has units of volts per degree. The Seebeck coefficient varies from material to material, of course, as do other physical constants such as thermal expansion coefficients or resistivity.

In order to understand the Seebeck effect we need to consider the behaviour of electrons in solids in some detail. However, we can gain an intuitive idea by assuming that the charge carriers behave like gas atoms. In this case, the charge carriers in the hot region will have a higher kinetic energy, and hence a higher velocity, then those in the cold region. Thus, in a temperature gradient, the net velocity of carriers at the hot end moving towards the cold end will be higher than the net velocity of the carriers at the cold end moving towards the hot end. In this situation more carriers will flow from the hot end towards the cold end than vice versa. This continues until the potential difference between the ends caused by this redistribution of the charge carriers is sufficient to impede further net transfer.

At this stage the material will be at equilibrium and the potential so set up is called the *Seebeck voltage*. One finds that for p-type semiconductors, the cold junction is positive with respect to the hot junction, and for n-type semiconductors, the cold junction is negative with respect to the hot junction. Thus a measurement of the sign of α will show if the material is n-type or p-type. For example, the non-stoichiometric forms of NiO, CoO and FeO will all yield positive values for α, indicating that conduction is by way of holes.

The main advantage of working with the Seebeck effect lies in the way in which the magnitude of the Seebeck coefficient varies with the number of charge carriers, and hence defects, present in the material. We will show below that α is large for small numbers of defects, and, in fact, the fewer the defects present, the larger α becomes. This is particularly useful for crystals such as NiO, which show only relatively small departures from stoichiometry.

In order to understand how this comes about, we need to consider the effect in terms of the thermodynamics of the system. Electrons (or holes) can be considered as chemically reactive species in the thermodynamic sense. Thus we can allot to them a chemical potential which is described by way of the electrochemical potential, $\bar{\mu}$, defined as

$$\bar{\mu} = \mu + ze\Phi \qquad (6.31)$$

where μ is the chemical potential of the system in the absence of an electrical potential, ze is the charge on the mobile species, and Φ is the electric potential in the neighbourhood of the carriers. In the present case, z will be equal to 1, as we are dealing only with electrons.

Turning now to a material subjected to a temperature gradient, as illustrated in Figure 6.7, when equilibrium is finally achieved we can write

$$\bar{\mu}_H = \bar{\mu}_C \tag{6.32}$$

i.e.

$$\mu_H + e\Phi_H = \mu_C + e\Phi_C \tag{6.33}$$

hence

$$e(\Phi_H - \Phi_C) = \mu_C - \mu_H \tag{6.34}$$

Setting the chemical potential μ equal to the Gibbs' free energy per particle, g, we can write

$$\mu = g = h - Ts \tag{6.35}$$

where h and s represent the enthalpy and entropy of the mobile charge carriers. Substituting from equation (6.35) into equation (6.34), we can write

$$e(\Phi_H - \Phi_C) = h_C - T_C s_C - (h_H - T_H s_H) \tag{6.36}$$

i.e.

$$e(\Phi_H - \Phi_C) = h_C - h_H + T_H s_H - T_C s_C \tag{6.37}$$

We now assume that the entropy of the hot and cold ends of the material can, to a reasonable approximation, be taken as equal, as can the enthalpies. Thus if we write $h_C = h_H = h$, and $s_C = s_H = s$, and these quantities are substituted into equation (6.37), it is found that

$$e(\Phi_H - \Phi_C) = (T_H - T_C)s \tag{6.38}$$

i.e.

$$(\Phi_H - \Phi_C)/(T_H - T_C) = s/e \tag{6.39}$$

However, we have defined α in equation (6.27):

$$+ \alpha = (\Phi_H - \Phi_C)/(T_H - T_C) \tag{6.40}$$

so that, for electrons

$$\alpha = - s/e \tag{6.41}$$

and for holes

$$\alpha = + s/e \tag{6.42}$$

We should note that the thermodynamic equations are not restricted to hopping semiconductors only, but apply to all materials with mobile charge carriers. Hence equations (6.41) and (6.42) apply to electronic conductors of all types. However, we can readily make these equations specific to hopping materials in the following way.

The entropy terms can be considered to be due to two parts, the arrangement of the charge carriers, which gives rise to the configurational entropy, and the vibration of the charge carriers due to thermal energy, which gives rise to the vibrational entropy. We can label these components S_c for configurational entropy and S_v for vibrational entropy. The value of S_v is difficult to calculate, but for a material with localized electrons, it is possible to estimate S_c by using the Boltzmann formula. This is identical to the calculation of the configurational entropy of point defects in a crystal as set out in Chapter 1. Thus we can write

$$S_c = k \ln [c!/(c-n)!n!] \qquad (6.43)$$

where S_c is the configurational entropy of n particles arranged on c available sites, and k is Boltzmann's constant. Using Stirling's approximation

$$\ln n! = n \ln n - n \qquad (6.44)$$

to simplify this expression, we can write

$$S_c = k[c \ln c - n \ln n - (c-n) \ln (c-n)] \qquad (6.45)$$

The entropy per particle, s_c, is given by dS_c/dn, which is

$$s_c = k \ln [(c-n)/n] \qquad (6.46)$$

However, as we have defined n/c as ϕ in the previous section, it is possible to write

$$s_c = k \ln [(1-\phi)/\phi] \qquad (6.47)$$

so that, from equations (6.41) and (6.42)

$$\alpha = -\{k \ln [(1-\phi)/\phi] + S_v\}/e \qquad (6.48)$$

for electrons and

$$\alpha = +\{k \ln [(1-\phi)/\phi] + S_v\}/e \qquad (6.49)$$

for holes. In general S_v is much smaller than $\ln [(1-\phi)/\phi]$ and as the only temperature variation will come into this equation from the S_v term, α will be approximately independent of temperature. To a good approximation, equations (6.48) and (6.49) are of the form

$$\alpha = \pm k[\ln (n_0/n_d) + A]/e \qquad (6.50)$$

where A is a constant, n_0 is the number of cation sites and n_d is the number of defects. This therefore confirms the statement made earlier that the value of the Seebeck coefficient will be largest for lower defect populations. If we take non-stoichiometric $Ni_{1-x}O$ as an example, n_0 would be the number of Ni^{2+} cation sites in the material, n_d the number of Ni^{3+} cations present and the value of α should increase as the composition of the $\approx NiO$ approaches $NiO_{1.000}$. Some data for the closely related material $Li_xNi_{1-x}O$ is presented in Figure 6.8 to illustrate the validity of equation (6.50).

Equations (6.48) and (6.49) also indicate that the value of the Seebeck coefficient will be large and negative if ϕ is close to zero, and large and positive

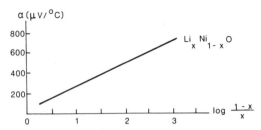

Figure 6.8 Variation of the Seebeck coefficient with defect concentration in $Li_xNi_{1-x}O$ crystals. The x-axis is a plot of $\log[(1-x)/x]$, which is equivalent to $\log[n_0/n_d]$ in equation (6.50). In this representation the number of defects decreases from left to right along the x-axis.

if ϕ is close to unity. Thus α will vary with stoichiometry, as did electronic conductivity. It is interesting to sketch out, in a qualitative way, this variation for a non-stoichiometric phase. As an example let us consider a non-stoichiometric oxide AO_2 which is fairly readily reduced to form AO_{2-x}, and which passes through the phases A_2O_3 and AO during the course of this reduction.

The sequence of events which occurs during the reduction has been described in the previous section when we considered the electrical conductivity of such a phase. Reference to this discussion shows that initial reduction will populate our AO_{2-x} crystal with a few A^{3+} ions which will give rise to n-type semiconduction. The value of α will therefore be large and negative. This value will fall as the number of defects increases, in accordance with our earlier analysis. Turning to A_2O_3, a slight degree of oxidation will introduce into the A_2O_{3+x} phase a small number of A^{4+} ions in a matrix of A^{3+} ions. This will lead to p-type semiconductivity and a large positive value for α. Continued oxidation will cause this value to fall as the number of A^{4+} centres increases. The way in which α is expected to vary over the composition range from AO_2

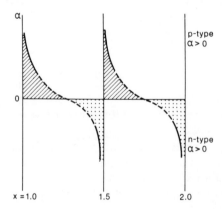

Figure 6.9 Schematic illustration of the expected variation of the Seebeck coefficient, α, with composition for a non-stoichiometric oxide system MO_x, where x can take values between 1.0 and 2.0.

to A_2O_3 is shown in Figure 6.9, where the region between the high-α and low-α regions has been extrapolated from the values near to the end compositions. This indicates a change from n-type to p-type behaviour at a composition of $AO_{1.75}$ and a value of zero for α.

Clearly, a similar situation will hold as we span the composition range between A_2O_3 and AO, with α taking a value of zero at a composition of $AO_{1.25}$. Moreover, the value of α will change drastically from large and positive to large and negative as we pass through the stoichiometric position at A_2O_3. One would therefore expect this effect to be observable in any phase which has a composition range either side of a stoichiometric composition, be it AO, A_2O_3 or AO_2. Experimental results from a wide variety of non-stoichiometric phases confirm the general trends that have been described.

Before leaving this topic it is worth pointing out that the Seebeck effect is reversible. That is, if a potential is applied across a material which contains a population of mobile charge carriers, some degree of cooling will be noticed at one end of the sample and some degree of heating at the other. For materials with large numbers of charge carriers the equations derived above suggest that the amount of heating and cooling so produced will be small. However, in non-stoichiometric materials with small defect populations this effect can become quite appreciable. Such materials therefore find uses in thermoelectric heating or cooling devices.

6.6 Classification of electronic conductors

6.6.1 *Band theory descriptions*

It is not the aim of this book to deal with band theory, which is covered in textbooks of solid state physics. In this section, therefore, we simply present a brief and qualitative account of the theory and how semiconducting transition metal oxides such as NiO or FeO are described in terms of band theory.

We can summarize the principal results of the theory by stating that the outer electrons in a solid, that is, the electrons which are of greatest importance from the point of view of electronic properties, are not constrained to discrete energy levels as in an atom, but occupy bands of allowed energies, which are separated from one another by bands of forbidden energies. The electron wave functions are considered to spread throughout the crystal and the electrons are described as being delocalized. The existence of energy bands leads to the prediction of a number of different types of material, depending upon the number of electrons in a band and the width of the energy bands themselves. The major types of electrical material are defined in terms of band theory in the following way.

(i) The number of electrons is such that at 0 K all the bands are either totally filled or completely empty, and at higher temperatures, thermal energy is considerably less than the energy gap between the top of the highest band of filled energies and the bottom of the next band of allowed energies. If the material contains no impurities that complicate matters, it will be an *insulator*. The highest *filled* band of energies is called the *valence band* and the first empty band of allowed energies above the valence band is called the *conduction band*. The energy difference between the top of the valence band and the bottom of the conduction band is called the *band gap*.

(ii) If the conditions in (i) apply, except that thermal energy is rather close to the band gap energy, some electrons will be transferred from the top of the valence band to the bottom of the conduction band at higher temperatures. The electrons in the conduction band will give rise to some degree of electronic conduction, of course. Rather surprisingly, we will also find that some contribution will also come from an equal number of positive charge carriers. These are called *positive holes*, or simply *holes*, and are usually regarded as positive counterparts to electrons moving in the valence band. Each time an electron is removed from the full valence band to the conduction band two mobile charge carriers are therefore created, an electron and a hole. Such materials are known as *intrinsic semiconductors*. At 0 K they will be insulators, but they will have an increasing conductivity as the temperature increases.

(iii) If a material of the type described in (ii) has a band gap similar to or less than the thermal energy, the number of charge carriers in each band becomes high and the material is classed as a *degenerate semiconductor*.

(iv) The conditions in (i) apply but the material contains an appreciable number of impurities which may have been added intentionally or not. The impurities are considered to act as *donors*, donating electrons to the conduction band, or as *acceptors*, accepting electrons from the valence band, which is equivalent to donating mobile holes to the valence band. Such materials are called *extrinsic semiconductors*. When donors are the main impurities present in the crystals, the conduction is mainly by way of electrons and the material is called an *n-type semiconductor*. Similarly, if acceptors are the major impurities present conduction is mainly by way of holes and the material is called a *p-type semiconductor*. If the donors and acceptors are present in equal numbers the material is said to be a *compensated semiconductor*. At 0 K these materials are insulators, and it is difficult in practice to distinguish between compensated and intrinsic semiconductors. When all of the impurities are fully ionized, so that either all the donor levels have lost an electron or all the acceptor levels have gained an electron, the *exhaustion range* has been reached.

(v) When there are insufficient electrons to fill the highest band, the valence band becomes indistinguishable from the conduction band. The material is said to be a *metal*, and the energy of the most energetic electrons in the partially filled band is termed the *Fermi level*. In a crystalline metal the Fermi level possesses a complex shape and is best called the *Fermi surface*.

(vi) In situations where the bottom of the $(n + 1)$th band lies at an

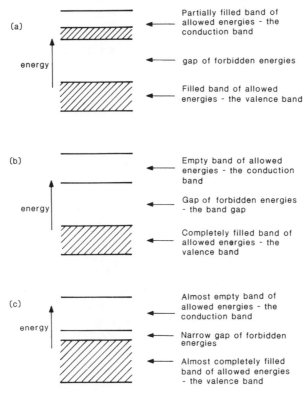

Figure 6.10 Schematic representation of the energy band scheme for (*a*), metals; (*b*), semi-conductors; (*c*), insulators.

energetically lower level than the top of the *n*th band, if the *n*th band is full, electrons will spill over into the bottom of the empty band until the Fermi surface intersects both sets of bands. Holes and electrons now coexist even at 0 K. This type of material is called a *semi-metal.*

These concepts are frequently summarized in the way shown in Figure 6.10. It is vital to remember, however, that such diagrams are enormously simplified and must be used with caution. The real band structure of a crystal is a complex three-dimensional shape, and a correct interpretation of even the more elementary aspects of electronic conduction in metals needs a more sophisticated approach than that given in this latter figure. Despite this limitation, we will utilize the simple diagrams of Figure 6.10 so as not to obscure the principles that we need to illustrate in this section, which concern defects in materials.

6.6.2 *Point defects in band-theory terms*

In our previous discussions of point defects we have concentrated upon an ionic description of the parent structure. Ionic solids can often best be regarded as insulators, as the electrons are localized on the ions making up these materials. Hence we expect the band picture appropriate to an insulator to be applicable. We now need to superimpose upon this the contribution of the point defects, which can be any of those defects that we have previously mentioned. The way in which this contribution is usually taken into account is to treat the defects as impurities and use the terminology set out in the previous section. Thus defects are treated as donors, which donate electrons to the conduction band to produce n-type semiconductors, or acceptors, which create holes in the valence band to produce p-type semiconductors.

Consider the case of interstitial cations. When interstitials are added to a crystal they are added as neutral atoms. However, in an ionic solid they are usually considered to be ionized. This means that the interstitials lose some electrons. In band-theory terms, each interstitial is represented as a donor level and any liberated electrons are placed in the conduction band, leading to n-type semiconductivity. This is shown schematically in Figure 6.11. If we concentrate upon interstitial anions, once again these must be added as neutral atoms. In an ionic crystal the neutral state is considered to be less stable than the ionized state, and in the case of non-metallic atoms they gain electrons to achieve this. Thus anions are represented as acceptor levels just above the valence band in energy. When the atom is ionized, electrons are gained from the valence band to leave behind mobile holes for conduction as shown in Figure 6.11.

The same sort of considerations will apply to vacancies. For instance, an anion vacancy gives rise to a set of localized levels just below the lower edge of the conduction band. If a neutral atom has been removed from the crystal, leaving behind one or more electrons, these may be trapped at the vacancy to produce a neutral defect, or placed in the conduction band if the vacancy is ionized, to cause n-type semiconductivity. A cation vacancy will be opposite to this in behaviour. Hence removal of a neutral metal atom from a material will involve removal of a cation plus the correct number of electrons. These will have to be taken from the valence band, which is the only source of electrons

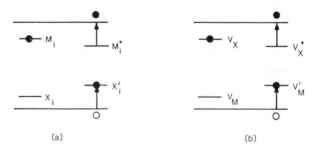

(a)   (b)

Figure 6.11 Schematic representation of the band structure of an insulating material containing point defects. Each defect is associated with an energy level, drawn as a short line. In (a) the defects are neutral and ionized anion and cation interstitials, and in (b), anion and cation vacancies. All defects are labelled in accordance with the Kröger–Vink notation.

that we have available. Cation vacancies will therefore be represented as acceptor levels situated near to the valence band. Both of these latter defects are shown in Figure 6.11.

The preceding discussion is detailed enough for our present needs, as we can see, for instance, that Zn interstitials in ZnO will be represented as donor levels, while cation vacancies in FeO will be represented as acceptor levels, in accord with our earlier descriptions. Only two other points need to be mentioned. Firstly, the amount of ionization of donor and acceptor levels, and hence their influence on electric properties, will vary with the energy of the centre. Hence one can talk about *deep energy levels* which have little effect on properties, and *shallow levels*, which have a larger effect. Secondly, it should be remembered that we are still considering a band model for the mobile charge carriers, so that the wave functions of the electrons and holes which take part in electrical conduction are delocalized and spread throughout the crystal.

6.6.3 *Ionic materials with localized electrons*

In the case of an ionic model the charge carriers are localized about their atomic centres. Charge movement is brought about by the jump of an electron from one such site or defect to an adjacent one, as we considered in detail above. This can occur only if (i) the latter is empty and (ii) if the electron acquires sufficient activation energy for the transfer. Such materials are known as *hopping semiconductors* or as *hoppers*, and the conduction process is activated, as we discussed earlier in this chapter. In such a model, the wave functions of the mobile charge carriers are localized and not spread throughout the crystal.

6.6.4 *Intermediate cases*

In some cases the electrons are neither free nor completely bound. These materials are called *narrow band* materials. If the electronic movement is only slightly impeded due to relatively weak interactions with the lattice, one describes the electron and its interaction as being a *large polaron*. The interaction takes the form of a small lattice polarization or deformation about the electron that moves along with the charge carrier. This is manifested in the apparent effective mass of the electron which is greater than its mass in the absence of such interactions, so that a large polaron moves in a band but has a slightly enhanced mass.

If the interaction is quite strong, one deals with a *small polaron*, consisting of a charge carrier surrounded locally by a strongly deformed lattice. The deformation tends to move with the carrier, in consequence of which its freedom to manoeuvre is considerably impeded. At low temperatures small polarons exhibit band-like properties, and they are able to tunnel slowly through the crystal as if in a band, but they possess a large effective mass which impedes their mobility considerably. As the temperature increases the band narrows, until the electronic behaviour of the material becomes indistinguishable from that of localized hopping-type materials.

6.6.5 Some examples

It would take us too far from the main direction of this book to list the electrical properties of many non-stoichiometric phases, but it is useful to make some brief comments here. Non-stoichiometric materials can be listed which span the whole range of electrical behaviour already mentioned. Leaving to one side the insulating materials discussed in Chapters 4 and 5, many sulphides, carbides and nitrides are good conductors of electricity and are well described in band-theory terms. This is also true of the oxides TiO_x and VO_x, where x takes values between approximately 0.9 to 1.3, and many of the so-called tungsten bronzes, which have compositions represented by the formula M_xWO_3, with x taking values between 0 and 1. The cation deficient monoxides of the $3d$ metals Fe, Co and Ni are well understood in terms of hopping semiconductivity. There is good evidence that conductivity is by way of polarons in slightly reduced WO_3, $SrTiO_3$ and TiO_2.

It must be borne in mind when looking at such generalizations that the electronic behaviour of many materials is quite complex. Thus it would be reasonable to assume that the electronic properties will vary greatly with both temperature and composition for materials best described in band-theory terms, as well as for hopping semiconductors. Each individual compound, therefore, must be treated on merit. For those who wish to take this interesting topic further, starting points are listed in the following section, Supplementary Reading.

6.7 Supplementary reading

There is no compact source of information on the material covered in this chapter, but related aspects, particularly for oxides, will be found in *Defects and Transport in Oxides*, eds. M.S. Seltzer and R.J. Jaffee, Plenum, New York (1975): *The Chemistry of the Solid State*, ed. C.N.R. Rao, Marcel Dekker, New York (1974); P. Kofstad, *Nonstoichiometry. Diffusion and Electrical Conduction in Binary Metal Oxides*, Wiley-Interscience, New York (1972); J.B. Goodenough, in *Progress in Solid State Chemistry*, Vol. 5, ed. H. Reiss, Pergamon (1972).

The band theory of solids is covered at a fundamental level in *Introduction to Solid State Physics*, C. Kittel, 5th edn., Wiley, New York (1975).

A very readable review article covering electron transport of in solids is D. Adler, in *Treatise on Solid State Chemistry*, Vol. 2, *Defects in Solids*, ed. N.B. Hannay, Plenum, New York (1975).

Electrical conductivity in oxides is discussed clearly with self-test questions by Ø. Johansen and P. Kofstad, *J. Mater. Ed.* **7** (1985) 909.

7 Colour centres

7.1 Introduction

In the previous chapter we discussed one of the most important aspects of non-stoichiometry in materials, the introduction of electronic defects into crystals. In this, we concentrated on the way in which electrons were transported through the structure and considered phenomena related to electronic conductivity. The electrons were often regarded as being trapped at atom sites, but not so strongly that activated movement was out of the question. In this chapter we draw attention to the fact that trapped electrons can have a profound effect upon the optical properties of the solids, and as such have a great potential for use in information storage devices.

Optical properties are associated, in general, with transparent materials, and so we will be considering those compounds that are principally composed of atoms that do not normally take more than one valence. These materials are insulators, typical examples being MgO, Al_2O_3 and ZrO_2. Trapped electrons in these originally colourless materials can sometimes cause their host structures to become deeply coloured, and the colours displayed by many natural semi-precious gemstones, such as amethyst and smoky quartz, arise in this way.

In the early studies of materials coloured by trapped electrons the sources of the colours were not known and the colour production was considered to be due to some uncharacterized *colour centres* in the material. The name has stuck and is now in common use as the generic noun for all related centres which produce colour in originally colourless materials by way of trapped electrons or their positive counterparts, electron holes. This chapter is therefore concerned with colour centres, and especially the relationship between colour centres and the stoichiometry of the host crystal that contains these defects.

7.2 The F-centre

Research in Germany, reported in 1938, indicated that exposure of alkali halide crystals to x-rays caused them to become brightly coloured. This colour was attributed to the formation of defects that were given the name

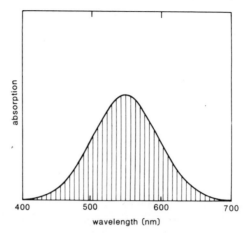

Figure 7.1 A typical bell-shaped absorption curve due to F-centres in KCl. Curves for other F-centres are similar in shape but displaced to other wavelengths.

Farbzentrum (= colour centre). These defects are now referred to by the briefer title of *F-centres*. Measurement of the absorption spectra of these crystals reveals a more or less bell-shaped curve of the type shown in Figure 7.1. Room-temperature data for colour centres in alkali halide materials are collected in Table 7.1.

Since then it has been found that many different types of high-energy radiation, including ultraviolet light, x-rays, γ-rays and neutrons, will cause these F-centres to form. The efficiency of the radiation with respect to F-centre production varies greatly, and x-rays, for example, tend to produce F-centres only in the surface layers of the crystal, while the more penetrating γ-rays give a uniform distribution of F-centres throughout the bulk of the material. One significant fact is that, regardless of the type of radiation used, the colour produced in any particular crystal is always the same. Thus, F-centres in NaCl are always an orange-brown colour and in KCl a violet colour, regardless of the method of F-centre production.

Table 7.1 Alkali metal halide F-centres

Compound	Fluoride		Chloride		Bromide	
	λ_{max}*	Colour	λ_{max}	Colour	λ_{max}	Colour
Li	224	—	388	Yellow-green	459	Orange
Na	344	—	459	Orange	539	Purple
K	459	Orange	563	Violet	620	Blue-green
Rb	—	—	620	Blue-green	689	Blue-green

*λ in nm.

An understanding of the true nature of F-centres has involved the correlation of a number of experimental results and the use of a variety of techniques. The first indirect experimental observation of interest was that at the same time that F-centres were produced in a crystal, its density fell. This shows that we must, if nothing else, be introducing vacancies into the structure. This poses an interesting problem which led to an insight into point defect chemistry which is not, strictly speaking, to do with colour centres themselves but is well worth mentioning.

The fact is that these vacancies cannot be created directly within the body of the crystal because the radiation used is usually not energetic enough to do this. The problem, therefore, is how to account for the diffusion of the vacancies into the crystal. The diffusion coefficients of both anion and cation vacancies are very low at room temperature, but the colour centres spread considerable distances into the crystals; a movement easily measured with an optical microscope. The difficulty was eventually explained by the suggestion that anion and cation vacancy pairs are the diffusing entities. This is reasonable to us as we have already noted that anion and cation vacancies carry opposite effective charges and hence are likely to associate in pairs. However, at the time this concept was put forward it was a novel idea, and formed the first suggestion that such vacancy pairs could exist. Calculations showed that the enthalpy of migration for a vacancy pair in alkali halide crystals is about $30 \, kJ \, mol^{-1}$ as compared to about $90 \, kJ \, mol^{-1}$ for a cation vacancy and $200 \, kJ \, mol^{-1}$ for an anion vacancy, supporting the contention.

Despite this success, which seemed able to account for the penetration of colour centres into a crystal, there was still no explanation for the origin of the colour. This is because we know that Schottky defects exist in alkali halides and hence that vacancy pairs will also exist in these materials, but alkali halide crystals are not coloured under normal circumstances. So vacancy pairs themselves cannot be the colour centres.

There are, it turns out, other ways in which we can produce F-centres in alkali halide crystals apart from using ionizing radiation. The first of these involves heating the crystals at high temperatures in the vapour of the alkali metal itself. In a similar way, if we grow crystals of an alkali metal in an atmosphere that contains an excess of alkali metal, colour centres again occur. It is also notable that the exact metal does not matter so long as it is an alkali metal. That is, if we heat a crystal of KCl in an atomsphere of Na vapour the typical KCl purple F-centres are formed, and not the orange NaCl colour centres. A third way of introducing F-centres into alkali metal crystals is to pass an electric current through heated samples and electrolyse them. In this case the typical F-centre colour is seen to move into the crystal from the cathode region. Once again, the colour depends upon the crystal being electrolysed and not the exact nature of the cathode.

These observations tell us that the centres are associated with defects in the crystal structure rather than the exact elements which constitute the compound. Moreover, experiments such as heating crystals in a metal vapour are reminiscent of some of the methods used to produce non-stoichiometric phases. What would we expect if this were happening? Consider KCl, for example. If we heat a crystal of KCl in an alkali metal vapour and a little is incorporated into the crystal structure it will occupy normal cation sites, as we know that Frenkel defects are not favoured in this compound. We can write the reaction thus:

$$A(g) \rightarrow A'_K + V^{\cdot}_{Cl} \qquad (7.1)$$

where A stands for the alkali metal added to the KCl crystal. Notice that the A'_K carries an effective negative charge because we have added a neutral metal atom to the system and that the vacancy carries the usual effective positive charge.

This state of affairs is not very likely, as alkali metals are more probably found as ions in the crystal. This is easily achieved by liberating the electron from the metal into the crystal. In the transition metal compounds that we discussed earlier this proved to be no problem, as the electron could sit at another cation site. Here, though, there is only one plausible site for the electron, and that is at the vacancy. The reaction suggested is

$$A'_K + V^{\cdot}_{Cl} \rightarrow A_K + V_{Cl} \qquad (7.2)$$

where the A is now in the normal 1+ state and the vacancy has an electron trapped at itself, that is:

$$V^{\cdot}_{Cl} + e' \rightarrow V_{Cl} \qquad (7.3)$$

Could this vacancy plus trapped electron be our F-centre? The trapped electron will undoubtedly absorb electromagnetic radiation if the energy of the photons corresponds to the trapping energy, and this could lead to an absorption spectrum of the type shown in Figure 7.1. We can check to see if this is reasonable by calculating the energy needed to free the electron using the same sort of very simple idea that we employed to calculate the energy of association of an anion and cation vacancy in Chapter 4. We start with a rather similar problem, that of an electron trapped at a proton, that is, with the hydrogen atom.

In a hydrogen atom the energy required to remove the electron completely from the atom, the ionization energy, is about 13.6 eV (1 eV corresponds to 1.6021×10^{-19} J). We can estimate this energy quite accurately by using the Bohr theory of the hydrogen atom, which we know gives good numerical answers for the energy levels of the electron in this situation. We will not reproduce the details of the Bohr model here, as it is to be found in many textbooks, one of which is listed in the Supplementary Reading section at the end of this chapter. The theory shows that the energy required to remove an electron completely from the nucleus is given by

$$E = -13.6/n^2 \, eV \qquad (7.4)$$

where the negative sign arises because the energy of the electron is taken as zero when it is free, and increasingly more negative as it approaches the nucleus. The escape energy is given by putting n equal to 1 in equation (7.4).

The formula was derived by assuming that the electron and the nucleus attracted each other in the same way as unit charges in a vacuum. This formula has already been given in Chapter 4, where we used it to calculate the interaction energy between point defects in an ionic crystal. To account for the effect of the crystal lattice, we simply diluted the attractive force by an amount equal to the dielectric constant of the crystal. If we use the same simple assumption here, then the formula given in equation (7.4) becomes

$$E = -13.6/\varepsilon n^2 \qquad (7.5)$$

where ε represents the dielectric constant of the crystal. As the magnitude of the dielectric constant for the alkali halides is about 5, we find that the energy to cause the electron to escape from the vacancy is about 2.7 eV. This corresponds exactly to the energy of the F-centre absorption band in NaCl. Of course we must not accord this figure such great significance. All it means, in view of the approximations involved, is that our idea that an F-centre consists of an anion vacancy plus a trapped electron is reasonable.

One more piece of evidence that we can call upon was not available to the earliest investigators. This is provided by the technique of electron spin resonance, which gives a measure of the number of unpaired electrons present in a solid. When this technique is applied to normal alkali metal crystals no unpaired electrons are found, of course. However, for crystals containing F-centres unpaired electrons are found, and in numbers equivalent to the number of colour centres present estimated by other means. This suggests that each F-centre contains one unpaired electron, as our model has proposed.

The present understanding is that an F-centre does indeed consist of a vacancy plus a trapped electron, as is shown in Figure 7.2. The defect is an

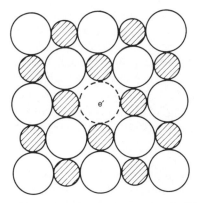

Figure 7.2 Schematic illustration of an F-centre in an alkali halide crystal. The colour centre consists of an electron trapped at a halide vacancy.

electron excess centre and arises because the crystal is slightly non-stoichiometric and contains a small excess of metal. Similar metal excess F-centres exist in compounds other than the alkali halides, of course. Fluorite, the mineral form of CaF_2, exists in a rare blue-purple form called Blue John. The coloration is caused by electron excess F-centres identical to those just described. In the same way non-stoichiometric ZnO, which contains small quantities of interstitial Zn atoms, is a bright red colour. The colour is again caused by the electron excess centres associated with electrons trapped at the interstitials.

7.3 Electron and hole centres

Colour centres of many types have been discovered since the original studies of F-centres. Many of these do not make crystals coloured, as they have adsorption maxima outside of the range of the visible spectrum. However, the term colour centre is used in a generic sense to refer to point defects or point defect clusters which have trapped electrons which can be liberated by radiation of energy similar to that of the visible region of the electromagnetic spectrum. The real structures of many of these trapping centres are not known with precision although possible structures have been described for many of them. We will not catalogue these here, but will refer to just two other interesting colour centres in this section. Many other examples will be found in the Supplementary Reading listed at the end of this chapter.

The first colour centre to mention is the F-centre formed in alkaline earth oxides such as CaO. For this defect to be neutral, two electrons must be trapped, as it will be recalled that an anion vacancy in such an oxide will have an effective charge of two units. If only one electron is trapped at such a vacancy it will still retain one unit of effective charge, and so can be written as an F^- centre. Such defects have been observed in CaO.

The second centre to mention is the one which gives rise to the colour in smoky quartz and amethyst. The crystal here is essentially silica, SiO_2, which contains a little Al as an impurity. As the aluminium is a trivalent ion which substitutes for silicon in the structure we need a method of preserving charge neutrality. In natural mineral crystals this is usually by way of incorporated hydrogen, which is present as H^+ in exactly the same amount as the Al^{3+}.

The colour centre giving rise to the smoky purple colour is formed when an electron is liberated from an $[AlO_4]^{5-}$ group by ionizing radiation and is trapped on one of the H^+ ions present. The reaction can be written as

$$[AlO_4]^{5-} + H^+ \rightarrow [AlO_4]^{4-} + H \tag{7.6}$$

It is seen that the $[AlO_4]$ group is now electron-deficient, and this type of colour centre is termed a hole centre. Although holes as such do not exist, it is conceptually easier to think of these as real entities, as we have already done in earlier chapters. The colour is then considered to arise from the trapping of a hole at the $[AlO_4]$ group. Other hole centres have been described in a variety of crystals.

7.4 Colour centres and information storage

7.4.1 *Information storage*

The ever-increasing use of electronic computers has generated a need for sophisticated means of information storage. Information for computers is measured in bits, with each bit having a value of 0 or 1. An average book is equivalent to about 10^6 to 10^7 bits, a figure estimated by assuming that the volume contains, let us say, 50 000 words of six letters each and each letter could be replaced by five bits, using the binary number system with A = 00001, B = 00010, C = 00011 and so on. Surprisingly, a single picture contains about the same number of bits. We can confirm this by supposing that the picture can be divided up into about 500 × 500 small areas, called pixels, without losing resolution, and that each pixel can have any one of 50 tone or greyness levels, so that each picture point needs 5 bits to define its intensity fully. The picture, at this level of resolution, needs 1.25×10^6 bits to define it. In principle, therefore, a single photographic negative can contain as much information as a book. This comparison illustrates the power of information storage via a memory plane.

The major advantage of the photographic film is that the high density of information is achieved by virtue of the fact that the $x–y$ location, or geographical position, of each pixel is also stored information, as well as the degree of darkening of each pixel. In addition, each pixel can be read or 'addressed' very rapidly indeed by way of scanned electron or light beams. The major drawbacks of photographic film as an information storage medium are that the films need chemical development and after this process the information is trapped in an irreversible fashion.

There is, therefore, considerable commercial interest in developing a film in which these drawbacks are eliminated. At the same time, such a film could be used for display screens and similar devices, as well as for information storage. In the following section we describe some research undertaken in the years around 1970 which used colour centres in CaF_2 for information storage. Although the work did not result in the development of a commercially successful product, the studies are interesting and provide an insight into the optical aspects of non-stoichiometry.

7.4.2 *Photochromic calcium fluoride*

From our discussion in the previous section we know that we are looking for a material which will markedly change colour when it is exposed to light. Such materials are called *photochromic*. Ideally, of course, we would like the change to be from transparent, or white, to opaque, or black. The degree of darkening should also, if possible, be directly proportional to the amount of light incident on the crystal.

The photochromic behaviour of inorganic crystalline materials usually results from the reversible transfer of optically excited electons from one type of trapping centre (let us call this A) to another type of trapping centre (B). When the electron is in site A, then the crystal will be clear, while when it is at site B the crystal will be dark. In order to make the changes controllable it is necessary to use light of one wavelength to transfer electrons from A to B and another wavelength to transfer them back again. This is illustrated schematically in Figure 7.3.

The transfer of electrons from a trapping centre under the influence of a light beam seems rather similar to the way in which F-centres colour the crystals in which they occur, and indeed this is so. Colour centres can be used to provide the sites needed for photochromic behaviour. One material which was extensively investigated for information storage use utilizing colour centres was CaF_2, which, we noted above, does occur naturally in a dark-blue form as well as in the normal colourless state.

The colour centres which form in CaF_2 to give this material a dark blue-purple coloration are normal F-centres. The colour, as we now know, is due to

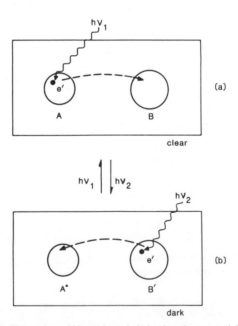

Figure 7.3 Schematic illustration of photochromic behaviour in a material such as $CaCl_2$. In (*a*) the crystal is clear and contains two unionized colour centres, A and B. Transfer of an electron from A to B using radiation of frequency v_1 gives ionized defects, A' and B', shown in (*b*). The crystal in this state is dark, but can be returned to the initial state by radiation of frequency v_2.

the absorption of light and the concurrent liberation of an electron from the halogen vacancy where it is trapped. Clearly the colour centre will remain 'ionized', that is, without its associated electron, only if some other defects can be provided which also act as electron traps. Earlier chapters suggest that a multivalent cation would provide a convenient alternative location. All that we need to be sure of is that the cation chosen will form a solid solution with the CaF_2 crystal, and that the trapping of the electron is not so strong as to prevent its return to the vacancy when required. A consideration of the crystal chemistry of CaF_2 showed that a rare-earth cation such as La would be a suitable second trap.

The initial stage in the preparation of photochromic CaF_2 involves 'doping' the crystals with LaF_3 to create one of the trapping sites. As we have discovered already, this will produce a non-stoichiometric phase. It is reasonable to assume that the large La ions will substitute for Ca and so occupy normal cation sites in the impure crystal. There are a number of ways in which the crystal can maintain charge neutrality, but in practice the crystals are grown under an atmosphere of HF and He. In these conditions the charge compensation is by way of F^- interstitials which appear to occupy an interstitial site next to a substituted La^{3+} cation, so that even at a local level the crystal remains neutral. The reaction is

$$LaF_3 \xrightarrow{\;CaF_2\;} La_{Ca}^{\cdot} + 2F_F + F_i' \qquad (7.7)$$

These crystals are 'water white' when grown.

The crystals are now made photochromic by the technique of heating in Ca metal vapour. This process is called *additive coloration*. It seems that the Ca metal atoms join the crystal surface and then increase the crystal volume because the F^- interstitials diffuse to the surface and occupy newly-formed anion sites. Of course the site rule means that for every Ca which joins the crystal we must create two F sites. The interstitials will only fill half of these, as we have only one interstitial per La. The other created site will remain as a vacancy in the lattice. This generates our second trapping site, which is nothing more than an F-centre in which the trapped electrons come from the Ca metal atoms as they are converted to ions. However, because of the La impurities, it is not exactly an ordinary F-centre. Instead of consisting of a halogen vacancy bounded by eight Ca ions, the new centre consists of a halogen vacancy bounded by seven Ca ions and one La ion. This site traps the two electrons provided by the Ca, one to neutralize the neighbouring $La^{\cdot}$ and one to neutralize the effective charge of $+1$ on the vacancy. The F-centre does not colour the crystal strongly but the crystals do take on a green coloration after this process. The reactions that take place are

$$Ca(g) \xrightarrow{\;CaF_2\;} Ca_{Ca}^{2'} + 2V_F^{\cdot} \qquad (7.8)$$

$$Ca_{Ca}^{2'} + 2V_F^{\cdot} \longrightarrow Ca_{Ca} + 2V_F \qquad (7.9)$$

$$\text{La}^{\cdot} + F' + 2V_F \longrightarrow F_F + (V_F' \text{La}^{\cdot}) \tag{7.10}$$

The last term in equation (7.10) represents the defect cluster composed of an anion vacancy containing two trapped electrons and an La^{3+} as one of the cations in the periphery of the vacancy.

It is these complex clusters that form the A sites in the clear material, although, as was noted above, the crystals are a greenish colour rather than completely transparent. The A sites are the thermally stable defects in the system and are the sites generated by the additive coloration process. The B sites are less complex and consist of La ions at Ca sites, that is, $\text{La}^{\cdot}$ units.

Photochromic switching occurs when an electron is transferred from an A site to a B site. When this happens the green colour diminishes and new absorption bands appear, turning the crystals a dark blue-black colour. The strength of the colours depends upon the concentration of A and B centres, of course, which in turn depends upon the actual conditions prevailing during crystal growth and subsequent heat treatment during the additive coloration process. In Figure 7.4 the absorption spectrum of the crystals in the initial state is shown. There is little absorption over the visible region, but a strong peak in the near ultraviolet, at about 390 nm. This is a good wavelength to use for ionization of the A sites, and if the crystal is illuminated with light near to this wavelength it will turn black as the electrons are transferred from A to B sites. This takes about two minutes at room temperature.

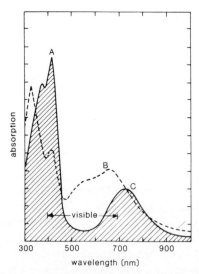

Figure 7.4 The absorption spectrum of CaF_2 doped with LaF_3. The continuous curve represents the clear state and the dotted curve the ionized dark state. Radiation corresponding to peak A is used to turn the crystal dark, i.e. to write information into the crystal. Peak B is used to reverse the process and turn the crystal clear, i.e. to erase information. Peak C is used to read information in the crystal. Radiation of energy corresponding to peak C does not alter the state of the crystal.

In order to erase the dark colour the electrons need to be transferred in the opposite direction. The absorption spectrum of the dark state is also shown in Figure 7.4. This reveals, besides a generally high degree of absorption across the visible region of the spectrum which gives rise to the darkening, a peak at about 680 nm. Light of wavelengths near to this peak will do the job for us. We thus have a system that we can darken or lighten at will. All that is now needed is some way of 'reading' the state of the crystal without altering it. This is achieved by using the absorption band which is present in the unswitched state at about 720 nm. It is found that irradiation at this wavelength does not alter the degree of darkening present, and so monitoring the intensity of this band provides a measure of the state of the system.

To summarize, we have developed an information storage film which does not need chemical processing and is reversible. We can write information into the film using light of wavelengths near to 380 nm, read information stored without changing the information using light of wavelengths near to 720 nm, and erase the information using light of about 680 nm. Moreover, the material does not degrade even after many cycles of writing and erasing, and these two processes are very efficient indeed.

This seems too good to be true and the immediate question is why the material is not being used if it is so good? There are two disadvantages. The first of these concerns the availability of lasers. Unfortunately there are no lasers which are ideal for the jobs of either writing, erasing or reading. The second disadvantage concerns the material itself. Firstly, the darkened state is thermally unstable and at normal temperatures lasts for only about a day. This state of affairs can be improved by replacing lanthanum by cerium, for example, but no perfectly stable darkened state has yet been perfected. Secondly, we are faced with the problem that in the non-stoichiometric system that we are working with only relatively low concentrations of defects can be introduced into the crystals. This means that we can only get useful changes in optical density by using thicker crystals, which increases the amount of material needed to store each bit of information. It is therefore necessary to try to find a way of greatly increasing the number of A and B defects present before a workable material can be produced. This is reminiscent of the situation that was encountered in the fast-ion conductors, where diffusion bottlenecks had to be overcome.

Although these disadvantages have so far meant that CaF_2 cannot be used for practical information storage devices, the idea of using colour centres is informative and reveals another of the interesting facets of non-stoichiometric compounds.

7.5 Supplementary reading

The best approachable source for an introduction to the topic of colour centres is K. Nassau, *The Physics and Chemistry of Colour*, Wiley-Interscience (1983), Chapter 9.

F

Bohr's theory of the hydrogen atom is clearly explained by H.C. Ohanian, *Physics*, Norton (1985), Chapter 41.

The information on photochromic CaF_2 is to be found in the scientific literature. Although these articles are rather advanced reading, they are worth studying to gain an insight into how the material was made and the efforts which were made in order to characterize it. The easiest starting points are D.L. Staebler and S.E. Schnatterly, *Phys. Rev. B* 3 (1971) 516; W. Phillips and C.R. Duncan, *Metall. Trans.* 2 (1971) 767.

8 Some thermodynamic considerations

8.1 Introduction

We saw in Chapter 6 that in order to maintain charge neutrality in non-stoichiometric crystals some sort of electronic defects could be involved in addition to atomic defects. It is clear that these electronic defects will impart to materials both interesting and technologically important electrical properties. It is also clear, though, that the existence of a non-stoichiometric phase will depend upon the partial pressure of the various components which make up the system. Thus $\approx$ FeO is unstable in air, the stable oxide being Fe_2O_3.

The relationship between composition and the partial pressure of the components is best understood in terms of the Gibbs' free energy of the particular system considered, and this brings us into the realms of thermodynamics. As can be imagined, the thermodynamics of non-stoichiometric compounds is complex and for many systems a satisfactory description is terms of thermodynamic formalism is still lacking. In this chapter therefore, we will indicate some areas where thermodynamics is helpful in the study of non-stoichiometric materials rather to give the flavour of the topic than to explore it in detail. We set the scene by considering a normal stoichiometric metal oxide and how it responds to changes in oxygen partial pressure. This has been chosen as more data are available for oxides, but the conclusions that are presented apply, of course, to all systems.

8.2 The equilibrium partial pressure of oxygen over an oxide

This is a topic which is described in most texts of chemical thermodynamics. Here we will cover only the bare bones of the matter, to save the reader having to consult alternative sources before coming to the later sections of this chapter. The approach will be illustrated by a specific example, the formation of silver oxide, and closely follows the treatment used in the thermodynamics text by Everett, cited in the Supplementary Reading section.

The formation reaction of silver oxide from silver metal and oxygen gas can be written as

$$4Ag + O_2 \rightleftharpoons 2Ag_2O \qquad (8.1)$$

155

The corresponding Gibbs free energy of a system containing silver, oxygen and silver oxide will be given by G:

$$G = n_{Ag}\mu_{Ag} + n_{O_2}\mu_{O_2} + n_{Ag_2O}\mu_{Ag_2O} \tag{8.2}$$

where n_A represents the number of moles of component A, and μ_A is its chemical potential. As the silver and silver oxide are pure solids, their chemical potentials, written μ_{Ag}^0 and $\mu_{Ag_2O}^0$, depend only on temperature. The chemical potential of oxygen, on the other hand, is given by

$$\mu_{O_2} = \mu_{O_2}^0 + RT \ln p_{O_2} \tag{8.3}$$

as oxygen is a gas in its standard state. In equation (8.3), $\mu^0{}_{O_2}$ is the standard chemical potential of oxygen gas at one atmosphere pressure, and p_{O_2} is the pressure of oxygen gas during the reaction, also measured in atmospheres.

If, during reaction, dn moles of oxygen are used, we form $2dn$ moles of silver oxide and consume $4dn$ moles of silver. The free energy of the system now becomes

$$(G + dG) = (n_{Ag} - 4dn)\mu_{Ag}^0 + (n_O - dn)\mu_{O_2} + (n_{Ag_2O} + 2dn)\mu_{Ag_2O}^0 \tag{8.4}$$

where dG is the change in free energy due to the reaction. Hence

$$dG = (2\mu_{Ag_2O}^0 - 4\mu_{Ag}^0 - \mu_{O_2})dn \tag{8.5}$$

and

$$dG/dn = 2\mu_{Ag_2O}^0 - 4\mu_{Ag}^0 - \mu_{O_2} \tag{8.6}$$

At equilibrium, dG/dn is zero, and substituting for μ_{O_2}, equation (8.6) becomes

$$2\mu_{Ag_2O}^0 - 4\mu_{Ag}^0 - \mu_O^0 = RT \ln p_{O_2} \tag{8.7}$$

The left-hand side of this equation consists of quantities which, at a given temperature, are all constant. In fact, this represents the standard free energy of formation of Ag_2O when one mole of oxygen is consumed. Thus we can write

$$\Delta G_f^0 = RT \ln p_{O_2} \tag{8.8}$$

Hence, at equilibrium, as ΔG_f^0 is a constant, the oxygen partial pressure, p_{O_2}, will also be constant. This pressure is called the *decomposition pressure* or *dissociation pressure* of the oxide, and depends only upon the temperature of the system.

What does this equation mean? Suppose we seal some silver metal and silver oxide in a closed silica ampoule, under a complete vacuum, and we heat the ampoule to a temperature T K. As we have no oxygen in the ampoule, some of the silver oxide will decompose, following equation (8.1), until the equilibrium decomposition pressure, as given by equations (8.7) or (8.8), is reached. If we raise or lower the temperature, either more silver oxide will decompose, or some silver will oxidize, until a new equilibrium decomposition pressure is achieved. Provided that we always have some silver and silver oxide in the tube, the oxygen pressure will depend only upon the temperature.

If we have an open system (such as a boat containing a mixture of silver and silver oxide) in a furnace, with a stream of gas flowing over the surface of the materials, the same reactions will occur. If the partial pressure of the oxygen in the gas stream is lower than the equilibrium partial pressure that would arise in the sealed-tube experiment, then some silver oxide will decompose. However, as the gas is flowing over the sample, the extra oxygen so generated will be swept away and the partial pressure of oxygen will not be augmented. Thus, more oxide will decompose, until finally all the oxide has been transformed and only silver remains. In the same way, if the partial pressure of oxygen in the gas stream is higher than the equilibrium decomposition pressure, some silver will oxidize. This will not deplete the partial pressure of oxygen over the metal plus oxide mixture, as the flowing gas stream will rapidly re-establish the original partial pressure. Hence, more silver will oxidize until ultimately the mixture has been converted totally to the oxide. The only conditions under which both silver and silver oxide will remain indefinitely in the boat occur when the partial pressure of oxygen in the gas stream corresponds exactly to the decomposition pressure at the temperature of the furnace.

This behaviour is often summarized on a graph of the type shown in Figure 8.1. On the vertical axis, $RT \ln p_{O_2}$ or ΔG_f^0 is usually plotted, instead of simply $\ln p_{O_2}$, and on the horizontal axis the temperature is recorded. Figure 8.1 contains rather a lot of information, so let us analyse it carefully. Firstly, we note that for every value of T, only one value of p_{O_2}, or $RT \ln p_{O_2}$, corresponds to the equilibrium value. At point A on Figure 8.1, for example, the pressure is given by $RT \ln p_{O_2}^A$ which corresponds only to the temperature T_A. If the oxygen partial pressure is lower than that represented by $RT \ln p_{O_2}^A$ then the silver oxide will decompose either until the pressure reaches $RT \ln p_{O_2}^A$, as in a closed system, or else until all the silver oxide has changed to

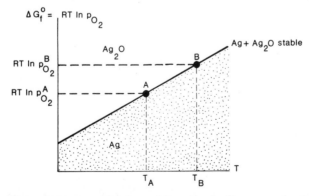

Figure 8.1 Equilibrium between a metal, Ag, and its oxide, Ag_2O, occurs only when external conditions of temperature and oxygen partial pressure given by the line passing through the points A and B exist over the metal/metal oxide mixture.

silver, as in an open system. Similarly, if we have an oxygen partial pressure greater than that represented by $RT \ln p_{O_2}^A$, then the silver plus silver oxide mixture will oxidize, either partially or completely, depending upon the volume of the container used for the experiment. Silver oxide and silver will only exist indefinitely at point A on the diagram, where both the temperature and the oxygen partial pressure have the values T_A and $p_{O_2}^A$. Of course, the same thing will happen at other temperatures such as at T_B, for example. Here we have taken T_B to be a greater temperature than T_A. In this case $RT \ln p_{O_2}^B$ will be greater than $RT \ln p_{O_2}^A$. Hence the equilibrium decomposition pressure will be greater, but otherwise the behaviour of our mixture will be the same as before.

Because the value of $RT \ln p_{O_2}$ is higher at higher temperatures, the series of points which represent the condition of equilibrium will lie on a line which slopes up towards the right, as shown on Figure 8.1. This equilibrium line is corresponds to the equation

$$\Delta G_f^0 = RT \ln p_{O_2} \qquad (8.8)$$

where ΔG_f^0 represents the standard free energy of formation of the silver oxide. Only when the oxygen partial pressure and temperature correspond to points on this line will the silver and silver oxide system remain unchanged. For combinations of temperature and oxygen partial pressure which yield points in the area above the line, only Ag_2O will be stable in an open system, and so this region is labelled Ag_2O. Similarly, for combinations of temperature and oxygen partial pressure represented by points below the line, only Ag will be stable in an open system, and consequently this region is labelled Ag.

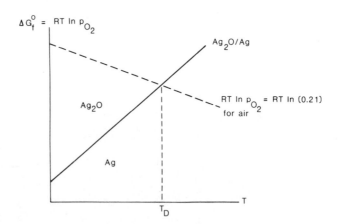

Figure 8.2 Graphical representation of the stability of an oxide in air. The line $RT \ln (0.21)$ represents the situation holding in air, which has an oxygen pressure of 0.21 atm, while the line Ag_2O/Ag represents the line marking equilibrium between silver and silver oxide. The lines cross at a temperature T_D, at which Ag metal and Ag_2O are in equilibrium in air. Heating the mixture in air to temperatures greater than T_D causes oxide decomposition.

A question of interest for scientists and engineers in many disciplines is whether a given oxide will be stable in air or not at a given temperature. Now this can readily be determined graphically from a diagram such as that of Figure 8.1, if we plot on it the data for air. As the vertical axis is $RT \ln p_{O_2}$ we need to plot this function for air, in which p_{O_2} is about 0.21 atm. It is readily found that a graph of $RT \ln (0.21)$ versus temperature is a straight line sloping from upper left to lower right (Figure 8.2). At the intersection of this line with

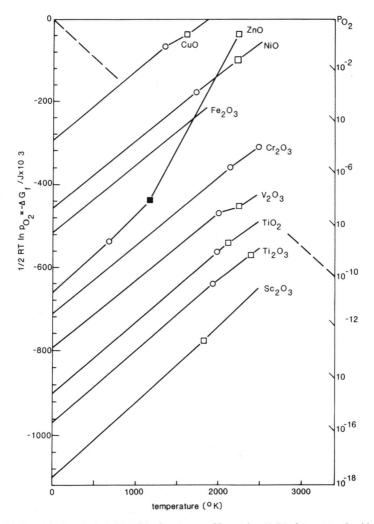

Figure 8.3 Graphical representation of the free energy of formation (ΔG_f) of some metal oxides as a function of temperature. The lines are all approximately parallel and slope upwards to the right. The broken line sloping downwards to the left is a plot of $RT \ln p_{O_2}$ for a pressure of 10^{-10} atm, and shows the temperature at which each of the oxides would decompose in an atmosphere with this partial pressure of oxygen. ○, □, melting points of the metallic element and the compound respectively; ■, boiling point of the element.

the equilibrium line for silver plus silver oxide, the metal and oxide will be in equilibrium with each other at the partial pressure of oxygen in air. This will occur at the temperature T_D, which is the temperature above which the oxide will decompose to silver in air. We can also see from Figure 8.2 that at temperatures higher than T_D, Ag metal will be stable, and at temperatures lower than T_D, Ag_2O will be stable.

The discussion we have just used is applicable to all oxides of course. If we plot the equilibrium lines for a number of metals we will end up with a chart containing a set of almost parallel straight lines, each one being similar to that drawn for the $Ag-Ag_2O$ system. Figure 8.3 shows this for some important metal–metal oxide systems. Such a chart is called an *Ellingham diagram.*

We can use these diagrams to determine the stabilities of oxides in various partial pressures of oxygen. Let us consider air. In this case we simply superimpose on the diagram the plot of $RT \ln(0.21)$ versus temperature and note where this line intersects the $RT \ln p_{O_2}$ versus temperature plot of the metal–metal oxide system of interest. The intersections of this line with the various other lines give the values of the decomposition temperatures of the oxides concerned. Lines representing other partial pressures of oxygen can be similarly plotted. On the right-hand side of Figure 8.3, values of p_{O_2} are listed, and we simply have to lay a rule along the graph from the origin to the requisite p_{O_2} mark to obtain the line corresponding to the value of p_{O_2} needed. In Figure 8.3 this is illustrated for a value of p_{O_2} of 10^{-10} atmospheres.

Although we have discussed oxides at length in this section, the same considerations apply to sulphides, chlorides and other compounds. Ellingham diagrams for these materials can be found in many textbooks of thermodynamics, mineralogy and metallurgy.

8.3 Variation of partial pressure with composition

From the foregoing discussion it is clear that if we have a closed system the equilibrium oxygen partial pressure will not depend upon how much Ag and Ag_2O is present. For example, if we have a sealed tube containing n moles of Ag_2O the pressure of oxygen will be independent of n. We can plot this graphically as in Figure 8.4.

A similar analysis to that presented above shows that for any pairs of oxides of a metal M, say M_2O_3 and MO_2, the same relationship must hold. Thus, if two oxides are in a sealed tube, the equilibrium partial pressure over them will be constant, regardless of how much of each oxide is present. This is the principle of action of *oxygen buffers.* To make an oxygen buffer, a large quantity of a metal plus its oxide, or in a multivalent system, a neighbouring pair of metal oxides such as Fe_3O_4, and Fe_2O_3, are placed in a closed system. Ultimately the partial pressure will reach the equilibrium pressure of the components chosen. This technique is widely used for generating known

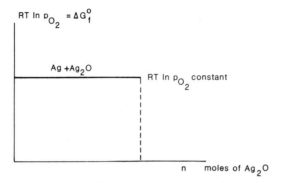

Figure 8.4 Graphical representation of the variation of the oxygen partial pressure, plotted as $RT \ln p_{O_2}$, versus composition, for a metal–metal oxide mixture, using the Ag–Ag_2O system as an example. The partial pressure will be constant for all proportions of the components of the mixture, and will depend only upon the temperature.

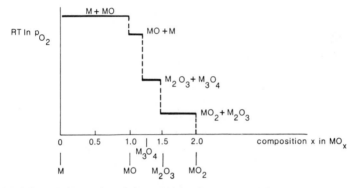

Figure 8.5 Schematic illustration of the variation of oxygen partial pressure across the metal–oxygen system of a transition metal M which forms stable oxides MO, M_3O_4, M_2O_3 and MO_2.

oxygen partial pressures, as the relevant free energy of formation data for most metal oxides is available in the literature.

If we have several oxides in a system, for example MO_2, M_2O_3, M_3O_4 and MO, the partial pressure of oxygen in the system will depend only upon temperature and the particular pair of oxides present. However, the pressure will change abruptly when we go from one oxide pair to another, say from MO/M_2O_3 to M_2O_3/MO_2. This is illustrated schematically in Figure 8.5, and some experimental data for the manganese–oxygen system confirming the theoretical reasoning is presented in Figure 8.6.

8.4 A thermodynamic definition of non-stoichiometry

The foregoing discussion allows us to define a non-stoichiometric phase in an alternative way to that given in Chapter 4. If we have two phases of fixed

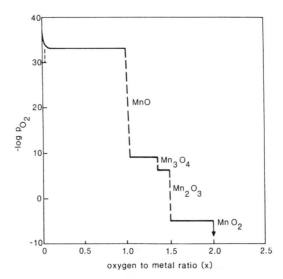

Figure 8.6 Variation of the equilibrium oxygen partial pressure with composition for the oxides of manganese. The full lines parallel to the composition axis represent two-phase regions and the dashed lines parallel to the $\log p_{O_2}$ axis represent single-phase regions.

composition coexisting at a temperature T, the partial molar free energy of each species is fixed. This means that the oxygen partial pressure over the two phases, if one or both is an oxide, will be fixed, as we have discussed above. The free energy or oxygen pressure in the system will not depend on the relative amounts of the two phases present, but only upon the temperature. Hence we can write

$$\mu_i = \mu_i(T) \tag{8.9}$$

where μ_i is the chemical potential of the ith phase, which is independent of the composition of the system, but depends upon temperature. Such a system is called *univariant*.

If, on the other hand, we have only one non-stoichiometric phase present, we would expect something different to happen. The easiest way to determine what this is, is to turn to the *phase rule*, which relates the number of components and phases in a system to the number of thermodynamic variables needed to define the state of equilibrium.

The phase rule was derived by J. Willard Gibbs in the latter half of the last century. It is extremely powerful, and also simple to use, provided that the terms in the equation are used in the strict sense that was specified by Gibbs himself. The 'rule' is usually written as

$$P + F = C + 2 \tag{8.10}$$

where P represents the number of phases in the system, and the word *phase* has a very specific connotation. Besides solids, the gaseous state is also a phase, for instance. In the same way, C, the number of components in the system, is precisely defined, and refers to the minimum number of chemical constituents that are needed to make up the compositions present in the system. The symbol F represents the minimum number of degrees of freedom that have to be specified in order to define the equilibrium completely. This is also called the *variance* of the system. (For an exact definition of all these terms, and examples in the use of the phase rule, refer to section 8.7, Supplementary Reading).

We will illustrate the use of the phase rule first of all for our silver oxide system, and then for a non-stoichiometric phase, to see how this leads to a thermodynamic definition of non-stoichiometry. In the case of silver oxide, equilibrium is achieved when three phases coexist: Ag_2O, Ag and O_2. The number of components in the system, in the precise way that we must use, is two, silver and oxygen, as we can make all of the phases present from just these two elements. Hence, substituting this information into equation (8.10), we find

$$3 + F = 2 + 2$$

i.e.

$$F = 1 \qquad (8.11)$$

What does this tell us? It tells us that under conditions where we have three phases in equilibrium we can only vary one thermodynamic parameter; all the others are dependent and fixed. We have already shown this to be true, of course. If we choose the temperature of our sealed ampoule, then the oxygen pressure is fixed. Similarly, if we choose pressure, only one temperature will allow the system to come to equilibrium. As temperature is usually the parameter most easily varied, we invariably choose this as our 'free' variable, making all the other parameters dependent. Thus the system has a variance of one; it is *univariant*, as we stated earlier.

Now consider a non-stoichiometric compound $\approx MO$ in equilibrium with the vapour phase. If we change the composition of the compound a little, we do not produce more of a second solid compound, but still have one solid in equilibrium with gas. The number of components in the system will be two, M and O_2, as in the case of silver oxide. Also, we know that the number of phases will be two, viz. oxide and gas, so we can write

$$P + F = C + 2$$

i.e.

$$2 + F = 2 + 2$$

thus

$$F = 2 \qquad (8.12)$$

The minimum number of degrees of freedom that have to be fixed to define the equilibrium is now two, and the system is said to be *bivariant*. So, in this case, if we know that we have $\approx MO$ and O_2 gas present in a sealed tube at a certain temperature we cannot be sure that we are at equilibrium. We must specify another variable to define the system completely. This could be pressure, but, once again, pressure is not as easy to manipulate as the composition of the oxide, so composition if often chosen in preference. Thus if we specify oxide $MO_{1.0980}$ in contact with O_2 gas at T K, the equilibrium oxygen partial pressure will be defined. Similarly, if we define the oxygen partial pressure and the temperature, the composition of the phase will have one precise value only.

Generally this behaviour can be expressed by an equation similar in form to equation (8.9). Composition and temperature are usually the variables which are selected. Hence, a bivariant system is written as

$$\mu_i = \mu_i(T, x) \tag{8.13}$$

where x represents the composition of the non-stoichiometric phase i.

Thus, if we have a system which appears to contain only one solid phase, and the behaviour is bivariant, then the phase can be termed a non-stoichiometric compound. This serves as thermodynamic definition of a non-stoichiometric compound to complement the structural one given in Chapter 4. To express this more explicitly, then, we can say that a chemical compound is a non-stoichiometric phase if, over a range of chemical composition, equilibrium is bivariant. This means that any temperature T, the chemical potentials of the components are continuous functions of the composition of the solid phase across its existence range, and indeed, this is the precise interpretation of equation (8.13).

Let us illustrate this behaviour for the Fe–O system. If we start by oxidizing iron metal in a sealed tube, the first pair of equilibrium products will be Fe coexisting with the lower composition range of $\approx$ FeO, say $FeO_{1.050}$. The reaction is

$$2Fe + 1.050\,O_2 \rightarrow 2FeO_{1.050} \tag{8.14}$$

The partial pressure in the tube, p'_{O_2}, is given by

$$\Delta G' = \Delta G_f = -RT \ln p'_{O_2} \tag{8.15}$$

where ΔG_f is the free energy of formation of $FeO_{1.050}$. In this case p'_{O_2} will be fixed and will depend only upon the temperature. Similarly, for the higher equilibria we can write, for $\approx$ FeO–Fe$_3$O$_4$:

$$6FeO_{1.150} + 0.275\,O_2 \rightarrow 2Fe_3O_4 \tag{8.16}$$

the partial pressure over the oxides being given by

$$\Delta G''_r = RT \ln p''_{O_2} \tag{8.17}$$

where $\Delta G''_r$ is the free energy of the reaction concerned. For the reaction of Fe_3O_4 and Fe_2O_3,

$$2Fe_3O_4 + \tfrac{1}{2}O_2 \rightarrow 3Fe_2O_3 \tag{8.18}$$

the partial pressure over the oxides is given by

$$\Delta G_r''' = RT \ln p_{O_2}''' \tag{8.19}$$

where $\Delta G_r'''$ is the free energy of the reaction given in equation (8.18). In all three reactions, (8.14), (8.16) and (8.18), the partial pressure of oxygen over the system will be fixed and depend only upon temperature. Thus the equilibrium situation will resemble that shown in Figure 8.5.

However, within the stoichiometry range of $\approx$ FeO, between the composition limits of approximately $FeO_{1.050}$ and $FeO_{1.150}$, only one solid phase will be present. The system will now be bivariant in behaviour, and the partial pressure of oxygen over the $\approx$ FeO will depend both on temperature and composition. In effect the pressure will vary continuously from p' to p'' as a function of the composition of the oxide. This is shown as a diagonal line in Figure 8.7. Changing the temperature will change the pressure versus composition variation, but it will still occur.

We can note here that the thermodynamic properties of the $\approx$ FeO phase are conveniently measured using a solid state cell employing stabilized zirconia as a solid electrolyte as described in Chapter 5. Using a technique such as this, one finds that at 1100 °C the equilibrium oxygen pressure changes from $10^{-13.2}$ to $10^{-10.7}$ atmospheres across the $\approx$ FeO phase range. At high temperatures, $\log p_{O_2}$ is very nearly a linear function of the composition, as we have shown in Figure 8.7. Below 1000 °C the dependence of chemical potential upon composition displays more complex features.

We close this section by noting that it is not always easy to be certain whether a system shows true bivariant behaviour or not. We return to this topic in Chapter 10.

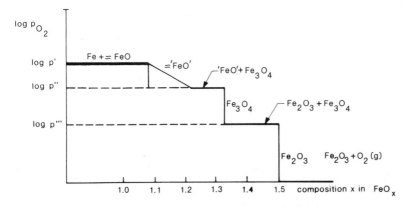

Figure 8.7 Schematic illustration of the variation of oxygen partial pressure, plotted as $\log p_{O_2}$ versus composition, x, for the iron–oxygen system. The heavy lines represent two oxides present at equilibrium and the diagonal line represents the non-stoichiometric $\approx$ FeO phase.

8.5 Electronic conductivity, composition and equilibrium partial pressure

As the composition or the impurity content of a non-stoichiometric phase changes, the number of defects present in the material fluctuates. If these defects give rise to observable physical properties such as electronic conductivity the magnitude of this property will also vary with composition. The fact that the composition and the equilibrium partial pressure of the components of a non-stoichiometric phase are inseparable allows one to measure the variation of the physical property with partial pressure and relate this to composition changes. It is this aspect of non-stoichiometric phases that we discuss in this section of Chapter 8. Once again, the principles that we wish to consider are general, but we will illustrate them by referring to the electronic conductivity of non-stoichiometric oxides.

8.5.1 Cation-deficient oxides

Let us commence our discussion with those oxides which accommodate their non-stoichiometry by way of cation vacancies, and take $\approx$ NiO as an initial example. In this oxide, as we have seen, the cation vacancies are balanced by compensating Ni^{3+} ions. We can write down a pseudo-chemical equation for the production of vacancies and Ni^{3+} ions following the guidelines set out in Chapter 4. This is

$$2Ni_{Ni} + \tfrac{1}{2}O_2(g) \rightleftharpoons O_O + V''_{Ni} + 2Ni^{\cdot}_{Ni} \qquad (8.20)$$

where V''_{Ni} represents a cation vacancy with two virtual negative charges and $Ni^{\cdot}_{Ni}$ is a Ni^{2+} cation on a normal Ni^{2+} site, but bearing an extra positive charge, that is, a Ni^{3+} ion. If there is some uncertainty that the charge is located on normal Ni^{2+} ions, we may prefer to write the equation as

$$\tfrac{1}{2}O_2(g) \rightleftharpoons O_O + V''_{Ni} + 2h^{\cdot} \qquad (8.21)$$

where $h^{\cdot}$ represents a positive hole, not trapped or located at any particular site in the lattice, but free to move through the valence band.

Having expressed the defect formation reaction in terms of a chemical equation, we can handle it by normal equilibrium thermodynamics. The equilibrium constant of reaction (8.20) is

$$K = [Ni^{\cdot}_{Ni}]^2[V''_{Ni}][O_O]/p_{O_2}^{1/2}[Ni_{Ni}]^2 \qquad (8.22)$$

where [ ] represent concentrations, p_{O_2} the oxygen partial pressure and K the equilibrium constant for the reaction. Now the values of $[O_O]$ and $[Ni_{Ni}]$ are essentially constant, as the change in stoichiometry is small. Hence we can assimilate them into a new constant, K_1, and write:

$$K_1 = [Ni^{\cdot}_{Ni}]^2[V''_{Ni}]/p_{O_2}^{1/2} \qquad (8.23)$$

We also know that for every vacancy in the crystal we have two Ni^{3+} ions, so that

$$2[V''_{Ni}] = [Ni^{\cdot}_{Ni}] \tag{8.24}$$

Hence

$$K_1 = [Ni^{\cdot}_{Ni}][Ni^{\cdot}_{Ni}][V'_{Ni}]/2p_{O_2}^{1/2} \tag{8.25}$$

so that the concentration of Ni^{3+} ion vacancies is given by

$$[Ni^{\cdot}_{Ni}] = \tfrac{1}{2}K_1 p_{O_2}^{1/6} \tag{8.26}$$

The concentration of Ni^{3+} ions is therefore proportional to the oxygen partial pressure to power 1/6. Similarly we could also say that the concentration of holes, $[h^{\cdot}]$, will also be proportional to the oxygen partial pressure to power 1/6, i.e.:

$$h^{\cdot} \propto p_{O_2}^{1/6} \tag{8.27}$$

The same sort of reasoning can be used with non-stoichiometric $\approx$ FeO. Let us consider this as a further illustration. Following the example above we can write

$$2Fe_{Fe} + \tfrac{1}{2}O_2(g) \rightleftharpoons O_O + V''_{Fe} + 2Fe^{\cdot}_{Fe} \tag{8.28}$$

or

$$\tfrac{1}{2}O_2(g) \rightleftharpoons O_O + V''_{Fe} + 2h^{\cdot} \tag{8.29}$$

Now as equations (8.28) and (8.29) are identical in form to equations (8.20) and (8.21), the subsequent analysis of the $\approx$ FeO case must be identical to that for $\approx$ NiO, and we will end up, once again, with equation (8.27).

In both examples given, the concentration of positive holes depends on the oxygen partial pressure in a definite way. It is apparent from our foregoing discussion that these oxides will be p-type semiconductors and, as it is reasonable to assume that the conductivity, σ, will be proportional to the concentration of holes, we can write

$$\sigma \propto p_{O_2}^{1/6} \tag{8.30}$$

and it is clear that the conductivity will *increase* with increasing oxygen pressure.

Before leaving this group of materials we will consider one more example, $\approx$ CoO. In both $\approx$ NiO and $\approx$ FeO we assumed that only one hole was localized on a cation at any one time. This is largely because Ni^{3+} and Fe^{3+} are realistic chemical species. In the case of $\approx$ CoO, though, Co^{4+} is also reasonable, so that instead of two holes sitting on separate cations to make two Co^{3+} ions, we could have them both on one cation to form a Co^{4+} ion. We can write this as

$$Co_{Co} + \tfrac{1}{2}O_2(g) \rightleftharpoons O_O + V''_{Co} + Co^{\cdot\cdot}_{Co} \tag{8.31}$$

or

$$\tfrac{1}{2}O_2(g) \rightleftharpoons O_O + V''_{Co} + (2h^{\cdot}) \tag{8.32}$$

where two holes are bracketed to show they are associated. Following through the analysis as in the cases of $\approx NiO$ and $\approx FeO$, we find

$$K = [O_O][V''_{Co}][Co^{\cdot\cdot}_{Co}]/[Co_{Co}]p_{O_2}^{1/2} \qquad (8.33)$$

i.e.

$$K_1 = [V''_{Co}][Co^{\cdot\cdot}_{Co}]/p_{O_2}^{1/2} \qquad (8.34)$$

because $[O_O]$ and $[Co_{Co}]$ are effectively constant and can be assimilated into K to form K_1. Also we know that $[V''_{Co}]$ is equal to $[Co^{\cdot\cdot}_{Co}]$ and hence

$$[Co^{\cdot\cdot}_{Co}] = [2h^{\cdot}] = K_1 p_{O_2}^{1/2} \qquad (8.35)$$

Again we expect p-type semiconductivity, which increases with oxygen pressure, but now the dependence is not proportional to power 1/6, but to power 1/4, i.e.

$$\sigma \propto p_{O_2}^{1/4} \qquad (8.36)$$

Experimentally it is not too difficult to measure conductivity as a function of oxygen pressure. If a graph of $\log \sigma$ versus $\log p_{O_2}$ is drawn, we should get a straight line graph, the slope of which, $+1/6$ or $+1/4$, will give us some information on whether the holes are free or associated in the structure. Figure 8.8 illustrates this for $\approx NiO$.

8.5.2 Oxygen-deficient oxides

In order to illustrate the behaviour of oxygen-deficient materials we will concentrate on those oxides which have interstitial cations in the structure and

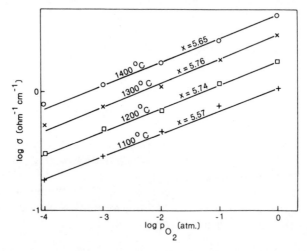

Figure 8.8 Conductivity plotted versus the logarithm of oxygen partial pressure for high-purity nickel oxide single crystals determined at a variety of temperatures. The slopes of the lines $1/x$, are close to, but not exactly equal to, 1/6. This data is redrawn from C.M. Osborn and R.W. Vest, *J. Phys. Chem. Solids* **32** (1971) 1131, 1343, 1353.

which are n-type semiconductors. Let us take ZnO as an example. When ZnO is heated in zinc vapour, we obtain a non-stoichiometric crystal containing excess zinc, $Zn_{1+x}O$. The reaction is

$$Zn(g) \xrightarrow{ZnO} Zn_i^{\cdot} + e' \qquad (8.37)$$

The equilibrium constant is given by

$$K = [Zn_i][e']/p_{Zn} \qquad (8.38)$$

where we assume that the interstitial zinc atoms are singly ionized to form Zn^{1+} ions. The concentration of these interstitials and also of the electrons is given by

$$[Zn_i^{\cdot}] = [e'] = Kp_{Zn}^{1/2} \qquad (8.39)$$

Thus the number of defects goes up as the vapour pressure of zinc metal goes up, and the semiconductivity will be proportional to the zinc vapour pressure, following the equation

$$\sigma \propto p_{Zn}^{1/2} \qquad (8.40)$$

We can treat the oxygen dependence similarly. In this case, heating ZnO in a vacuum will cause oxygen to be lost from the crystal, to again produce $Zn_{1+x}O$. Thus we can write the chemical reaction as

$$ZnO - O_o \rightleftharpoons Zn_i^{\cdot} + e' + \tfrac{1}{2}O_2(g) \qquad (8.41)$$

and the equilibrium constant as

$$K = [Zn_i^{\cdot}][e']p_{O_2}^{1/2}/[ZnO][O_o] \qquad (8.42)$$

If we follow previous procedures and amalgamate the concentrations of ZnO and lattice oxygen, which will be large and almost constant, into the equilibrium constant K to generate a new constant, K_1, we can rewrite equation (8.42) as

$$K_1 = [Zn_i][e']p_{O_2}^{1/2} \qquad (8.43)$$

Now equation (8.41) shows that

$$[Zn_i^{\cdot}] = [e'] \qquad (8.44)$$

so that

$$[Zn_i^{\cdot}] = [e'] = K_1 p_{O_2}^{-1/4} \qquad (8.45)$$

Hence both the number of defects and the electronic conductivity will be proportional to the oxygen partial pressure to power $-1/4$, i.e.:

$$\sigma \propto p_{O_2}^{-1/4} \qquad (8.46)$$

It is seen that the situation here is opposite to that encountered in the p-type

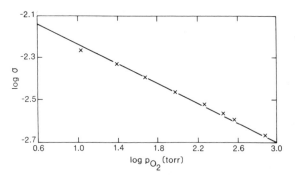

Figure 8.9 Classical data showing the conductivity of $\approx$ ZnO as a function of oxygen pressure at 650 °C as determined by Baumback and Wagner in 1933. The data is evidence that the Zn_i atoms are singly-ionized rather than doubly-ionized.

oxides, and now the conductivity will *decrease* as the oxygen partial pressure increases. During this process the composition of the material will approach the stoichiometric formula of $ZnO_{1.000}$. Some classical experimental data for $\approx$ ZnO, shown in Figure 8.9, indicate that equation (8.46) is obeyed fairly well, and gives confidence to the assumption originally contained in equation (8.41) that the interstitial Zn atoms are singly ionized and exist as Zn^{1+} entities in the structure.

8.5.3 Some further considerations

An essential consideration in each of the cases discussed has been the nature of the defects supposed to be present and their degree of ionization. In the last example considered, for instance, the zinc interstitials in ZnO might be thought to be doubly ionized. The chemical reaction expressing their formation from zinc vapour would then be of the form

$$Zn(g) = Zn_i^{\cdot\cdot} + 2e' \tag{8.47}$$

If the analysis is now followed through in the same way as before, we will clearly end up with a different relationship for the variation of electrical conductivity with partial pressure of zinc vapour to that set out in equation (8.40). In fact, we would find

$$\sigma = p_{Zn}^{-1/3} \tag{8.48}$$

The correct model can only be determined by resort to experimental data. As we have just noted, the electrical conductivity data in Figure 8.9 support the choice of singly-charged zinc interstitials in the present system.

The generation of interstitial ions is not the only way to produce oxygen-deficient materials. Let us consider the oxide TiO_2 in the crystalline form known by its mineralogical name of *rutile*. If rutile is heated in a vacuum at above about 1300 K it loses oxygen and becomes dark blue in colour. The composition is found to correspond to a chemical formula of TiO_{2-x}. There are two reasonable ways to account for the apparent oxygen deficiency. It is possible to suppose that oxygen vacancies have been generated in the structure. Thus we could write

$$O_O \rightleftharpoons V_O^{\cdot\cdot} + \tfrac{1}{2}O_2(g) + 2e' \qquad (8.49)$$

The electrons can be envisaged to be localized on separate Ti^{4+} ions to produce Ti^{3+} ions in the lattice, or both electrons could be localized on one Ti^{4+} ion to make it into a Ti^{2+} ion. Alternatively we could suppose that slightly reduced TiO_{2-x} contains interstitial metal atoms. We could set up defect equations to express all these three equilibria and complete a thermodynamic analysis to produce three contrasting equations for the relationship between the electronic conductivity and the oxygen partial pressure.

Ultimately it is necessary to resort to experiments to determine which of the possibilities is correct. In fact it is found that a model involving interstitial titanium atoms is correct, but even after only small degrees of reduction these interstitials are not distributed at random, but ordered in precise and regular arrays so that the idea of interstitial point defects in the way that we have used up to now becomes incorrect and quite misleading. It is therefore necessary to be cautious in the interpretation of data in terms of isolated or random point defects in a crystal structure. We discuss this matter, and the reasons why a cautious approach is necessary, in Chapter 9.

8.6 Brouwer diagrams

8.6.1 Introduction

Many of the important properties of materials, particularly in the field of electronic applications, are controlled by the defect concentrations in a crystal. In the preceding section we laid great stress upon the fact that for non-stoichiometric crystals electronic conductivity and defect populations are linked and can be closely controlled by manipulation of the surrounding atmosphere. Thus, if we want to specify the exact electronic conductivity of such a material we need to specify its chemical composition, the defect populations and types of defect present, the temperature of the crystal and the partial pressures of the constituents which make up the material. Of course, these are not all independent of one another, but a set of equations interrelating these variables does not convey an easily assimilated picture of the relationships which hold. More importantly, the way in which the properties of the material vary as the external conditions are changed remains obscure.

In these circumstances it is extremely helpful if the defect concentrations can be displayed graphically as a function of some variable such as partial pressure of one of the components. These diagrams can then show, in a lucid way, how the predominant defects change with the variables chosen, and allow the implications of the changes to be understood. It is, of course, nearly as difficult to draw graphs as it is to write down sets of simultaneous equations. However, a procedure introduced by G. Brouwer allows such figures to be constructed.

We start with defect reaction equations of the sort used earlier in this chapter. Consider, for example, the incorporation of oxygen into $\approx$ FeO, as we discussed in the previous section. If the holes so produced are not localized at Fe^{2+} sites we use equation (8.29), thus:

$$\tfrac{1}{2}O_2(g) \rightleftharpoons O_O + V''_{Fe} + 2h^{\cdot}$$

The equilibrium constant for this reaction, if we use more formal equations of chemical thermodynamics than those applied above, is given by

$$K = a(O_O) \cdot a(V''_{Fe}) \cdot a(h^{\cdot})^2 / p_{O_2}^{1/2} \tag{8.50}$$

where $a(i)$ represents the activity of the species i. If we assume defect concentrations are low and defect interactions are negligible, we revert to the same approximation employed earlier, viz.:

$$K' = [V''_{Fe}][h^{\cdot}]^2 / p_{O_2}^{1/2} \tag{8.51}$$

where square brackets represent concentrations, and the concentration of oxygen ions on oxygen sites $[O_O]$ has been taken as a constant and incorporated into K'. We will use this type of simplified equation to construct diagrams which show the way in which defect concentrations vary with the partial pressure of the components, and hence with composition. The procedure is best illustrated by way of examples, and these constitute the rest of this section.

8.6.2 Brouwer diagram for a non-stoichiometric phase $\approx MX$

For our first example let us suppose that only vacancies are important in the material, and interstitial defects do not play a significant role in the non-stoichiometric behaviour of the phase. We will furthermore suppose that the $\approx MX$ compound can have an existence range which spans both sides of the stoichiometric composition, $MX_{1.00}$. Naturally the stoichiometric composition will only be obtained when the number of vacancies on the cation sub-lattice is exactly equal to the number of vacancies on the anion sub-lattice. This is identical to the Schottky defect situation described in Chapter 1. In order to keep the example as simple as possible, we will also suppose that the electrons or holes in the non-stoichiometric compound are not trapped at the vacancies, but are free to move around by hopping or any other mechanism that we choose. In our non-stoichiometric phase $\approx MX$ we therefore have only four defects to consider, electrons, e', holes, $h^{\cdot}$, vacancies on metal sites V''_M and vacancies on anion sites, $V^{\cdot\cdot}_x$. Finally, we will assume that the most important gaseous component is X_2, as is the case in most oxides, halides and sulphides.

The equations analogous to (8.29) which describe the equilibrium between the species are the following.

(i) Creation and elimination of vacancies:

$$V''_M + V^{..}_X \rightleftharpoons \text{zero} \tag{8.52}$$

(ii) Creation and elimination of electronic defects:

$$e' + h^{.} \rightleftharpoons \text{zero} \tag{8.53}$$

(iii) Changes of composition:

$$\tfrac{1}{2}X_2 \rightleftharpoons X_X + V''_M + 2h^{.} \tag{8.54}$$

(iv) Maintenance of electrical neutrality:

$$2[V''_M] + [e'] \rightleftharpoons 2[V^{..}_X] + [h^{.}] \tag{8.55}$$

The equilibrium constants for the first three of these reactions can be written down respectively as

$$K_1 = [V''_M][V^{..}_X] \tag{8.56}$$

$$K_2 = [e'][h^{.}] \tag{8.57}$$

$$K_3 = [V''_M][h^{.}]^2/p^{1/2}_{X_2} \tag{8.58}$$

To obtain an equilibrium equation for reaction equation (8.55) which is not too complicated for our purposes, we assume that one of the defect species on each side in equation (8.55) is by far the most important. For the moment, then, let us suppose that there are far more cation vacancies than electrons, and that this is balanced by a much larger concentration of holes than anion vacancies. Equation (8.55) then simplifies to

$$2[V''_M] = [h^{.}] \tag{8.59}$$

Recall that we have already used this equation when discussing the behaviour of non-stoichiometric NiO, earlier in this chapter.

We can now substitute from equation (8.59) into equations (8.56), (8.57) and (8.58) to obtain relationships between the partial pressure of X_2 and the defect concentrations present in the material. Starting with equation (8.58):

$$[V''_M](2[V''_M])^2 = K_3 p^{1/2}_{X_2}$$

$$4[V''_M]^3 = K_3 p^{1/2}_{X_2}$$

$$8[V''_M]^3 = 2K_3 p^{1/2}_{X_2}$$

so that

$$[V''_M] = \tfrac{1}{2}(2K_3)^{1/3} p^{1/6}_{X_2} \tag{8.60}$$

and, in addition

$$\tfrac{1}{2}[h^{\cdot}][h^{\cdot}]^2 = K_3 p_{X_2}^{1/2}$$

$$[h^{\cdot}]^3 = 2K_3 p_{X_2}^{1/2}$$

so that

$$[h^{\cdot}] = (2K_3)^{1/3} p_{X_2}^{1/6} \tag{8.61}$$

From equations (8.57) and (8.61) we can also write

$$[e'] = K_2/[h^{\cdot}]$$

so that

$$[e'] = K_2(2K_3)^{-1/3} p_{X_2}^{-1/6} \tag{8.62}$$

and from (8.56)

$$[V_X^{\cdot\cdot}] = K_1/[V_M'']$$

so that

$$[V_X^{\cdot\cdot}] = 2K_1(2K_3)^{-1/3} p_{X_2}^{-1/6} \tag{8.63}$$

It is not immediately clear that this has helped a great deal. However, Brouwer proposed that by taking logarithms of each side of the four equations (8.60), (8.61), (8.62) and (8.63), linear relationships resulted which could be drawn as straight lines on a graph of the logarithm of the concentration of the species as a function of the logarithm of the partial pressure of X_2. The slope of these lines will be $\pm 1/6$ in the present case and are drawn as such in Figure 8.10.

Let us review the information contained in Figure 8.10. Firstly, we see that the concentration of holes is much higher than the concentration of electrons, as we chose, so that the material will be a p-type semiconductor. In addition, the number of holes will increase as the partial pressure of the gaseous X_2 component increases, and the number of electrons will fall. Thus, the p-type behaviour will be enhanced. Similarly, the number of metal vacancies will increase as the partial pressure increases while the number of anion vacancies fall. As these numbers were equal on the left-hand side of the diagram, we must have stoichiometric $MX_{1.0}$ present. As the partial pressure of X_2 increases, therefore, we tend to an anion excess compound, which is due to an increase in the number of vacancies on metal sites.

All this is implicit in the equations that we have used and assumptions that we have made, but the graphical representation makes the variation come alive. However, it is important to remember that the equations and assumptions are likely to be valid only over a restricted temperature regime, and so the diagram may well be a reasonable approximation only for a narrow temperature range or even one specific temperature.

To summarize, the procedure adopted in constructing a Brouwer diagram

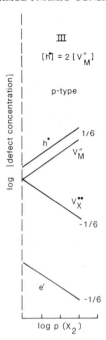

Figure 8.10 Partial Brouwer diagram for a phase $\approx$ MX in the region where the partial pressure of the gaseous component X_2 is high. The logarithm of the defect concentrations, $\log[h^{\cdot}]$, $\log[V_M'']$. $\log[V_X^{\cdot\cdot}]$ and $\log[e']$ are plotted along the y-axis, versus the logarithm of the partial pressure of X_2 along the x-axis. The slopes of each of the lines is $1/6$, and the material will be a p-type semiconductor over the whole region covered.

is to assume that one of the concentrations on each side of the neutrality equation is so dominant that the others can be neglected and then to write down the resulting linear relationships. The information given in the diagram will then be valid for the range of values of the temperature and partial pressure of X_2 for which the assumption is reasonable.

To return to the present example, let us now consider what will happen if the partial pressure of X_2 is less than the range shown on Figure 8.10. We see that we are moving towards a more stoichiometric composition for the $\approx$ MX phase. Now stoichiometric crystals tend to be insulators, and hence it is natural to suppose that in these circumstances the formation of vacancies on cation and anion sites is far more important than the creation of electrons and holes. In terms of our equations this is equivalent to the assumption that K_1 is a lot greater than K_2. From equations (8.56) and (8.57) we then see that the concentration of vacancies $[V_X^{\cdot\cdot}]$ will be much greater than the concentration of electrons, $[e']$. A more reasonable approximation for equation (8.49) is therefore

$$2[V_M''] = 2[V_X^{\cdot\cdot}] \tag{8.64}$$

instead of (8.59).

If we now work through the previous analysis we find that

$$[V_M''] = [V_X^{\cdot\cdot}] = K_1^{1/2} \qquad (8.65)$$

that is, the concentration of vacancies is not dependent upon the partial pressure of X_2. We have a stoichiometric phase containing Schottky defects, as we expect. Also

$$[h^{\cdot}] = (K_3^{1/2}/K_1)p_{X_2}^{1/4} \qquad (8.66)$$

and

$$[e'] = K_2 K_1 K_3^{-1/2} p_{X_2}^{-1/4} \qquad (8.67)$$

We can plot these on the log concentration versus $\log p_X$ graph, as before, to produce the result shown in Figure 8.11.

In Figure 8.11, we have included the equations which represent the major assumptions in both regions covered, which are now labelled II and III, and have also noted the slope of the lines which describe the variation of defect concentration with partial pressure of X_2. In the middle region we have chosen the number of cation and anion vacancies to be the same, and so the material will be strictly stoichiometric with a composition $MX_{1.0}$ over all of this range.

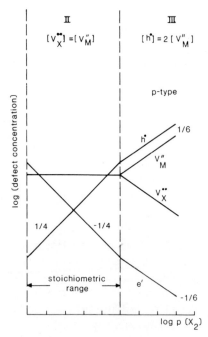

Figure 8.11 Partial Brouwer diagram for a phase $\approx$ MX, extended to lower partial pressure than applicable in Figure 8.10. The fraction beside the lines represents the slope of the lines. As the numbers of cation and anion vacancies is equal in the region to the left of the dotted line, the material in this part of the diagram has a composition $MX_{1.0}$.

In addition we have shown the division between the different regions as abrupt ones, which may not be true, of course. An intermediate region between the two might well be represented by an electroneutrality equation of the type

$$2[V_M''] = 2[V_X^{\cdot\cdot}] + [h^{\cdot}] \tag{8.68}$$

This will cause the sharp breaks in Figure 8.11 to be smoothed into curves, by interposing another region between those shown on the figure.

If we keep decreasing the partial pressure of X_2, we would expect the conditions to change again. In order not to make too abrupt a transition to this region of lower partial pressure we can imagine that we will go through an intermediate region corresponding to an electroneutrality equation

$$2[V_M''] + [e'] = [V_X^{\cdot\cdot}] \tag{8.69}$$

into a region where we have anion vacancies and electrons as the only important components of the electroneutrality equation. This equation can be written

$$[e'] = 2[V_X^{\cdot\cdot}] \tag{8.70}$$

and the region where it applies can be labelled region I. We can use equation (8.70) in a now familiar way to substitute into equations (8.56), (8.57) and (8.58). Because equation (8.70) is so similar in form to equation (8.59), we expect our equations for the concentrations $[e']$, $[h^{\cdot}]$, $[V_X^{\cdot\cdot}]$ and $[V_M'']$ to be similar to equations (8.61), (8.62), (8.63) and (8.64).

On re-examining these latter equations, we find that there is no equation relating $[V_X^{\cdot\cdot}]$ and $[e']$ to the partial pressure of X_2 analogous to equation (8.58). This equation is needed because of the components in equation (8.70), of course. It is, however, very easy to construct such an equation by writing

$$X_X \rightleftharpoons \tfrac{1}{2}X_2(g) + V_X^{\cdot\cdot} + 2e' \tag{8.71}$$

which is the equation corresponding to (8.54). We can then write

$$K_4 = [V_X^{\cdot\cdot}][e']^2 p_{X_2}^{1/2} \tag{8.72}$$

Actually, we are not introducing a new and arbitrary equilibrium constant, as the four equations (8.52)–(8.55) still suffice to define the system. This is because

$$K_4 = K_1 K_2^2 / K_3$$
$$= [V_M''][V_X^{\cdot\cdot}][e'][h^{\cdot}][e'][h^{\cdot}]p_{X_2}^{1/2} / [V_M''][h^{\cdot}]^2 \tag{8.73}$$

and we could simply have relied upon rearrangement of K_1, K_2 and K_3 for our answer. The advantage of writing an equation such as (8.71) is simply that it leads one to the correct form for the equilibrium constant rather more quickly than otherwise.

We can now substitute from equation (8.71) into (8.72) or (8.73) to obtain

$$[V_X^{\cdot\cdot}] = \tfrac{1}{2}(2K_1^2/K_3)^{1/3} p_{X_2}^{-1/6} \tag{8.74}$$

$$[e'] = (2K_1K_2^2/K_3)^{1/3}p_{X_2}^{-1/6} \qquad (8.75)$$

$$[h^{\cdot}] = \{K_2/(2K_1K_2^2/K_3)^{1/3}\}p_{X_2}^{1/6} \qquad (8.76)$$

$$[V_M''] = \{2K_1/(2K_1K_2^2/K_3)^{1/3}\}p_{X_2}^{1/6} \qquad (8.77)$$

These equations can now be plotted if we take logarithms of both sides. The result is shown in Figure 8.12, plotted as region I.

In this figure we see presented the whole Brouwer diagram for the system, and it is worth taking some time to see exactly what the figure contains in the way of information. Before doing this we should note that one change has been introduced in Figure 8.12 compared to Figures 8.10 and 8.11; the Schottky defect formation reaction equilibrium constant K_1 has been rewritten K_S, and the electronic defect formation reaction equilibrium constant K_2 by K_e, as these latter terms are the more usual ones found in the literature. Also, of course, there are now three regions corresponding to low, medium and high partial pressure of X_2 gas, and it is readily seen that the electron concentration starts high, in the n-type region I, and falls progressively, while the hole concentration starts low and ends high in the p-type region III. The way in which the other defects vary in concentration is also easily determined from

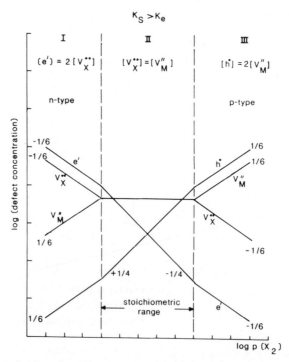

Figure 8.12 Complete Brouwer diagram for a phase $\approx$ MX for conditions of equilibrium specified by the electroneutrality equations shown in the three regions I, II and III.

the figure. This variation gives a number of clues about the way in which the properties of the phase change as the partial pressure of X_2 and the composition alter across the diagram. At the left, for example, there is high concentration of anion vacancies and so easy diffusion of anions is to be expected. Similarly, in region III we have a high concentration of cation vacancies which would be expected to enhance cation diffusion.

Our original assumption was that Schottky defect formation was preferred to the formation of elctronic defects, and we have stated this explicitly at the top of the figure. Indeed, as we have ignored Frenkel defect formation altogether we could have written $K_S > K_e \gg K_F$, where K_F would represent the equilibrium constant for a reaction which formed Frenkel defects in MX. The diagram as given in Figure 8.12 therefore contains all the assumptions made, as well as the way in which defect concentrations vary. Moreover, it is easy to see what changes will take place if the relative values of K_S and K_e are altered. Such information is contained in the equations that we have used, of course, but the graphical representation is able to present trends in a far more lucid fashion.

8.6.3 *An experimentally-determined Brouwer diagram: CdTe*

The diagram given in Figure 8.12 is, of course, schematic and it would be a straightforward exercise to construct a number of such diagrams for different situations. Rather than do this, it seems more useful to consider a real situation to illustrate some of the advantages of the Brouwer diagram method of expressing point defect equilibria. As semiconducting materials have been investigated in most detail we have chosen one of these, the phase CdTe.

CdTe is an important semiconductor device material. It has a structure similar to silicon, but with alternating layers of Cd and Te atoms instead of layers of Si atoms alone. In the presence of excess Cd the material is an *n*-type electronic semiconductor, while in the presence of excess Te it is a *p*-type electronic semiconductor. The Brouwer diagram is shown in Figure 8.13. The diagram shows, on the vertical axis, the concentrations of the defects per cubic centimetre. On the presssure axis, the units chosen are $\log K_r p_{Cd}$, which refers to the reaction

$$Cd(g) \rightleftharpoons Cd_i^{\cdot} + e' \tag{8.78}$$

The equilibrium constant for this reaction, K_r, is given by

$$K_r = [e'][Cd_i^{\cdot}]/p_{Cd} \tag{8.79}$$

What information does the figure contain? First of all we can see that there are five defects considered to be of importance, holes, $h^{\cdot}$, electrons, e', neutral cadmium interstitials, Cd_i^X, cadmium interstitials which have an effective charge of $+1$, $Cd_i^{\cdot}$, and vacancies on cadmium sites which also have a trapped hole associated with them, V'_{Cd}. The point defect equilibrium in region II is

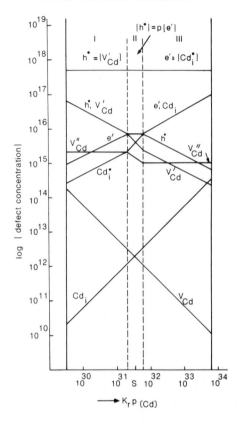

Figure 8.13 Brouwer diagram for the semiconductor compound CdTe, derived from experimentally determined values for the defect concentration and the partial pressure of Cd vapour.

considered less important than the electronic defects, so that this region is defined by the equation

$$[e'] = [h\,\dot{}] \tag{8.80}$$

In region I, corresponding to low partial pressures of Cd, the charge neutrality condition was taken as

$$[h\,\dot{}] = [V'_{Cd}] \tag{8.81}$$

leading to p-type conductivity and probable Cd diffusion via vacancies in the Cd sub-lattice. At high partial pressures of Cd, in region III, we have a charge neutrality equation

$$[e'] = [Cd_i\,\dot{}] \tag{8.82}$$

which gives n-type conductivity, and a high number of Cd interstitials. We see that stoichiometric $CdTe_{1.000}$ is formed when the number of Cd interstitials equals the number of Cd vacancies, that is, when

$$[Cd_i] = [V_{Cd}]$$

and

$$[Cd_i^\cdot] = [V_{Cd}'] \tag{8.83}$$

which occurs at a precisely defined pressure of Cd vapour, marked S on the x-axis of the figure.

The diagram could be complemented by a similar one for Te defects, of course. Would these be important in practice? That will depend upon the real magnitude of the equilibrium constants for the defect formation reactions compared to those utilized in Figure 8.13.

This very brief discussion of the CdTe diagram concludes these remarks on Brouwer diagrams. A great deal of literature is available concerning the production and use of the diagrams in a variety of situations. The references given in the following section should be consulted if more information is required.

8.7 Supplementary reading

The thermodynamics of solid equilibrium with gas atmospheres is considered in many textbooks of thermodynamics, and often, with useful examples, in textbooks of metallurgy and geology. A basic introduction to the subject is provided by D.H. Everett, *Introduction to the Study of Chemical Thermodynamics*, Longman, London (1959).

The thermodynamics of many metal–oxygen systems are especially well characterized in view of their industrial importance. Two very useful descriptions of gas-solid equilibria are T.B. Reed, *Free Energy of Formation of Binary Compounds: an atlas of charts for high-temperature chemical calculations*, MIT Press, Cambridge (1971); A. Muan, The effect of oxygen pressure on phase relations in oxide systems, *Am. J. Sci.* **256** (1958) 171–207.

A very readable account of oxide equilibrium, with self-test questions, is given by G.A. Smiernow and L. Twidwell, *J. Ed. Mod. Mater. Sci. Eng.* **1** (1979) 223.

The thermodynamics of non-stoichiometric compounds and the relationship between thermodynamics and structures has been covered at an advanced level, but clearly, by J.S. Anderson. See particularly J.S. Anderson, *Bull. Soc. Chem. France*, No 7 (1969) 2203; J.S. Anderson, in *Problems of Non-stoichiometry*, ed. A. Rabenau, North-Holland, Amsterdam (1970).

The use of the phase rule is clearly explained by E.G. Ehlers, *The Interpretation of Geological Phase Diagrams*, Freeman, San Francisco (1972). In this respect the original works of J. Willard Gibbs are also extremely interesting. They have been republished in two volumes by Dover Publications (1961).

Much experimental data concerning electronic conductivity in oxides and its variation with oxygen partial pressure will be found in P. Kofstad, *Non-Stoichiometry, Diffusion and Electrical Conductivity in Binary Metal Oxides*, Wiley-Interscience, New York (1972).

The original description of Brouwer diagrams is well worth consulting: G. Brouwer, *Philips Res. Reports* **9** (1954) 366.

More information, and a large number of examples, will be found in F.A. Kröger, *The Chemistry of Imperfect Crystals*, 2nd edn., North-Holland, Amsterdam (1974); W. van Gool, *Principles of Defect Chemistry of Crystalline Solids*, Academic Press, New York (1966).

Part 4 The structures of non-stoichiometric phases

9 Structural considerations

9.1 Introduction

In previous chapters we have discussed a defect chemistry which has allowed us to go some way towards describing non-stoichiometric compounds analytically. In this chapter we attempt to describe the real structures of these materials, and in the following chapter, which ends this book, we summarize how the ideas already expressed need to be adapted to be consistent with this experimental data.

Before embarking upon this account, it is necessary to mention experimental techniques. Our definition of a non-stoichiometric phase rests upon the concept of a structure persisting over a range of anion to cation ratios. Thus a key feature of any discussion of non-stoichiometry must be an evaluation of the resolution or discrimination of the technique which is used to characterize the structure and the ability of the technique to detect when a second phase appears. If the history of non-stoichiometric compounds is surveyed one is constantly finding that a compound once considered to possess a broad composition range is found instead to consist of a series of well ordered *microphases* simply because the resolution of the technique employed to characterize the structure has improved.

In all probability the technique which has done most to clarify the understanding of the structures of these materials is high-resolution electron microscopy. This is because the method is able to reveal the structure of disordered materials, which is difficult with other modes of structure analysis. High resolution electron microscopy will not be described here, but the additional sources listed in the Supplementary Reading section provide detailed information both on the technique itself and how it has clarified the problems addressed in this book.

9.2 Small departures from stoichiometry

There are a number of materials which show small but measurable departures from stoichiometry. The best known of these compounds are probably the transition metal monoxides typified by NiO and CoO, but a large number of other materials also fall into this class, and we can cite as examples ZnO, CdO,

185

Cu_2O, V_2O_3, VO_2 and NbO_2. In some of these phases the composition range spans both sides of the stoichiometric composition; NbO_2 for instance has a reported composition range of from $NbO_{1.9975}$ to $NbO_{2.003}$. In others, the materials have a composition range on only one side of the stoichiometric composition. In CoO, for example, the composition can range from approximately $Co_{0.99}O$ up to $Co_{1.00}O$, while in CdO the composition ranges from $Cd_{1.00}O$ to approximately $Cd_{1.0005}O$.

With all members of this group of compounds there is some uncertainty about the degree of non-stoichiometry tolerated, as the difficulties associated with measuring the existence range are considerable. Partly for this reason their defect structures are uncertain, and such phases are usually treated in the literature as containing isolated point defects. Thus NiO and CoO, both of which possess the NaCl structure when fully stoichiometric, are considered to accommodate composition changes by way of a population of vacancies on the normally occupied metal positions. In the case of CdO, which also has the NaCl structure, the metal excess is usually considered to be due to interstitial Cd atoms or ions. As the following section will show, such a simple description may not be correct.

9.3 Point defect clusters

Even when the departures from stoichiometry remain small, there is every reason for believing that the defect structures of non-stoichiometric phases are far more complex than a picture of isolated point defects would suggest. Probably the best documented example is that of $\approx$ FeO, wüstite. This oxide is stable above about 840 K, and in general never attains the composition $FeO_{1.0}$. Instead, as we have already noted in earlier chapters, the oxide exists over a composition range which varies from about $Fe_{0.89}O$ to $Fe_{0.96}O$ at approximately 1300 K and which broadens with increasing temperature. It therefore belongs to the same group as NiO and CoO described above, but possesses an appreciably larger composition range.

From a powder x-ray point of view, the structure of $\approx$ FeO is of the NaCl type. The classical picture of $\approx$ FeO is that the oxygen array is perfect and the non-stoichiometric composition is due to vacancies among the iron atoms as we discussed, for example, in Chapter 4. However this is a very poor approximation to the real structure. Single crystal x-ray photographs and electron diffraction patterns reveal the presence of extensive regions of diffuse scattering which are indicative of short-range order. Analysis of diffraction data has shown that isolated iron vacancies are not present at all, but instead small groups of atoms and vacancies aggregate into elements of new structure which are distributed throughout the wüstite matrix. Such regions are often termed *point defect clusters*, and although this term has some utility, it is misleading because point defects are not present. Some of the cluster arrangements so far characterized are shown in Figure 9.1. It is found that

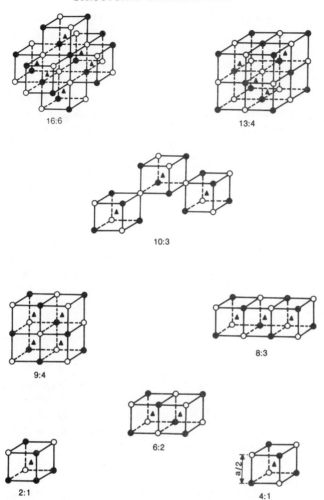

Figure 9.1 The proposed structures of clusters which are believed to occur in $Fe_{1-x}O$. The open circles represent Fe vacancies, the filled triangles Fe in tetrahedral coordination, and the filled circles Fe in octahedral environments. Oxygen atoms have been omitted for clarity. The numbers below each diagram represent the ratios of Fe vacancies to Fe in tetrahedral positions.

these clusters bear a strong resemblance to elements of Fe_3O_4, the next higher oxide to FeO. Structurally, therefore, the over-oxidized FeO is a partly ordered assembly of fragments of the Fe_3O_4 structure intergrown within the NaCl structure expected of an oxide of formula FeO. The fact that these clusters are more stable than equivalent point defect populations has been confirmed by calculation of lattice energies, a point we take up further in Chapter 10.

Although a good deal of experimental and theoretical study has been carried out on the defect structure of $\approx$ FeO, some aspects are still obscure and more work is needed. What is certain, however, is that a point defect model for the structure is not a very accurate reflection of the real situation which obtains in this oxide.

Few other systems with a composition MX have been as carefully investigated, but it is increasingly clear that many other non-stoichiometric phases behave in quite a similar fashion. From among the monoxides, $\approx TiO$ and $\approx VO$ have complex structures involving the ordering of 'vacancies' on both the metal and oxygen sites, so that the simplistic concept that these oxides contain point defect populations is incorrect. In the sulphide $\approx FeS$, which has the NiAs structure, a large number of superstructures can exist, many of which have yet to be properly characterized. Nominally these phases contain ordered Fe vacancies and a perfect sulphur sub-lattice, as in wüstite, but, in reality, the 'vacancies' are part of the structure of the compound and are not, in any structural sense, accurately regarded as point defects.

The list of materials in this section could be extended greatly, and from among the sulphides alone $\approx ZrS$, the chromium sulphides near CrS in composition, $\approx TiS$, and the phases between digenite, Cu_2S, and bornite, Cu_5FeS_4, come to mind. All of these materials, with stoichiometries close to MX, show complex superlattices or diffuse scattering on diffraction patterns, and although the non-stoichiometry of these materials is frequently described in terms of point defects, there is little doubt that more elaborate studies will reveal that clusters of structure occur in which the notional point defects are assimilated completely.

Another large group of compounds which also show pronounced clustering or ordering of 'point defects' are those with structures related to the fluorite type with ideal composition MX_2. The composition range of these phases can be either above or below the composition of the ideal fluorite structure. For a good many years it has been known that in these compounds it is the non-metal lattice which is the seat of the non-stoichiometric variation and the formulae of these phases are thus MX_{2+x} or MX_{2-x}.

In the MX_{2-x} phases, clusters, which to a first approximation can be thought of as consisting of pairs of oxygen vacancies, form. These 'vacancy pairs' are aligned along $\langle 111 \rangle$ directions in the cubic fluorite cell. Ordering of strings of these clusters in a regular fashion generates a homologous series of phases which have compositions expressed by the series formula M_nO_{2n-2}. Thus, although these phases are often considered to contain variable populations of anion vacancies, such descriptions are misleading.

In the MX_{2+x} phases we nominally have interstitial anions present. Once gain these are not random point defects, but arranged in clusters. Indeed, one of the earliest cluster geometries to be understood is that of the so-called Willis 2:2:2 cluster in the fluorite structure oxide UO_{2+x}. The structure of this cluster is shown in Figure 9.2, from which it can be appreciated that the point defects notionally present have been incorporated into a localized restructured region.

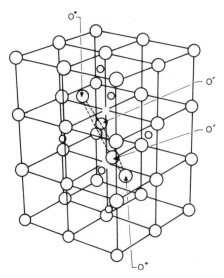

Figure 9.2 The structure of the 2:2:2 cluster in UO_{2+x}. Two interstitial oxygen atoms, labelled O′, interact with two normal oxygen atoms which are thus displaced to interstitial sites to become the O″ atoms, thereby creating two vacancies, O_V, on normal sites. The site is called a 2:2:2 cluster because it is made up of $2O′, 2O″$ and $2O_V$ entities.

As with the MX phases, so with the MX_2 compounds, and a large number of other examples of non-stoichiometric oxides containing defect clusters could be cited, including the rare earth oxides, the stabilized zirconias and the bismuth oxides. Regrettably it would require considerable space to consider these phases, even in summary, and this would be beyond the scope and purpose of the present chapter. Once again we can stress, though, that in all of these compounds that have been investigated structurally in sufficient detail it is found that the notional point defects have assimilated into the structure. Clusters of 'point defects' form and are fully integrated into the host matrix. It is the concentration of these clusters which varies and which accounts for the composition ranges of the phases observed experimentally.

9.4 Interpolation

Interpolation is quite a widespread means of changing the composition of a compound and encompasses a very broad range of materials from clathrates and intercalation compounds to interstitial alloys. Some aspects of these phases were mentioned in Chapter 4, and we will not dwell on them at length in this survey. For completeness, though, we will draw attention to one or two examples of particular interest.

The tungsten bronzes are perhaps the classical examples of such phases. In Chapter 4 we met the perovskite tungsten bronzes, which have a formula

M_xWO_3, and which are formed by the reaction of alkali metals with WO_3. The WO_3 structure has large cages available between corner-linked WO_6 octahedra, and these cages can be filled with metal atoms. The degree of filling, x, is variable and the compounds have wide composition ranges. In the example quoted earlier, Li_xWO_3, x can vary continuously between the values of 0 and 0.5. For the closely related Na_xWO_3, x can vary between the limits of 0 and 0.11, and then from 0.41 to 0.95. Many other examples of these compounds will be found in the literature.

In addition to the perovskite-structure tungsten bronzes, two other tungsten bronze types exist which make use of interpolation to vary their compositions. These are both made up of corner-linked WO_6 octahedra, and in them the WO_6 octahedra are re-arranged to form either pentagonal, square, or hexagonal tunnels, as shown in Figure 9.3. Variable filling of these tunnels by metal atoms gives rise to wide stoichiometry ranges. In the tetragonal

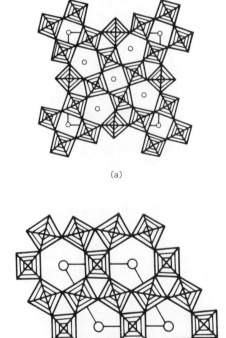

(a)

(b)

Figure 9.3 The tetragonal tungsten bronze structure (*a*), and the hexagonal tungsten bronze structure (*b*). The shaded squares represent WO_6 octahedra, which are linked to form pentagonal, square and hexagonal tunnels. These are able to contain a variable population of metal atoms, shown as open circles.

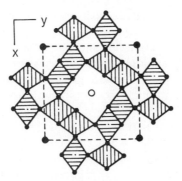

Figure 9.4 The hollandite structure. The framework of the structure is composed of continuous chains of metal–oxygen octahedra, shown as shaded diamonds, which are linked so as to form square tunnels. A variable population of cations, shown as filled or open circles, can occupy these tunnels to give the materials significant composition ranges.

tungsten bronze phases, shown in Figure 9.3(a), both the pentagonal and square tunnels are partly filled. These phases are typified by the compounds Pb_xWO_3 and Sn_xWO_3, in which x can take values of from approximately 0.16 to 0.26. In the hexagonal tungsten bronzes, shown in Figure 9.3(b), it is the hexagonal tunnels which are partly occupied. The composition of the hexagonal tungsten bronze phase K_xWO_3 extends from $x = 0.19$ to $x = 0.33$ in this way.

We end this section by briefly mentioning the *hollandite* structures. These are rather similar, structurally, to the hexagonal tungsten bronzes in that they contain tunnels that can be filled with large metal atoms in variable proportions. In these materials, however, the parent compound is TiO_2, and the tunnels are square in cross-section, and made up of chains of edge-shared TiO_6 octahedra, as shown in Figure 9.4. Hollandites have received considerable attention in recent years for possible use in the storage of radioactive materials. For this purpose the radioactive metal ions, which are normally the larger ions in the Periodic Table, enter the tunnels, where they are trapped by enclosing the hollandite matrix in an inert shell.

9.5 Planar faults and boundaries

In the preceding examples of non-stoichiometry, the structural ways in which the crystals are able to accommodate changes in the atomic ratios were very much related to the concept of point defects. In the remainder of this chapter we will consider systems in which quite different mechanisms are used, often employing some form of planar boundary to accomplish the same task.

In general, if we insert a planar boundary into a crystal, its stoichiometry will change, the most abrupt example being the crystal surface itself. If such boundaries are introduced in variable numbers and distributed at random, we will generate a crystal with a variable composition. If the faults order, a series of new phases will be generated. Each phase will be characterized by the separation between the ordered planar boundaries, which are no longer faults or defects, but integral structure elements. In addition, each phase will have a fixed composition, although the formula may involve large integers, and will differ in composition from its neighbours by a small but definite amount. There is another general feature which accompanies the introduction of planar faults into crystals. New coordination polyhedra are often created in the vicinity of the discontinuity that are not present in the parent structure. These may provide sites for novel chemical reactions or introduce significant changes in the physical properties compared to those of the parent structure.

Examples of such phases have been known for a number of years in the mineralogical field, where the terms *intergrowth phases* and *polysynthetic twinned phases* have been applied to describe the structures. It is only more recently, however, that planar defects have been generally recognized as important in non-stoichiometric compounds. A large number of such systems is now known, and as experimental techniques for the structural characterization of non-stoichiometric materials improve, more will undoubtedly be uncovered.

9.5.1 Crystallographic shear phases

The phenomenon of *crystallographic shear* (*CS*) seems to be important mainly in the transition metal oxides WO_3, MoO_3, Nb_2O_5 and TiO_2, and provides a mechanism for altering the anion to cation ratio in these materials without either changing the shape of the anion coordination polyhedra of the metal atoms significantly or introducing point defects. In the oxides in which *CS* occurs, the metal coordination polyhedron is an usually an octahedron of oxygen atoms. These are linked by corners or edges and corners to form rather open structures. On reduction, either by removing oxygen directly or by doping with atoms of lower valence, the open structure collapses to produce crystallographic shear planes, (*CS* planes) which have the effect of eliminating oxygen. The lower oxides so produced are known as *crystallographic shear phases*.

Geometrically, the simplest compounds that we can use to illustrate this process occur in the tungsten–oxygen system. The structure of the parent oxide, WO_3, is shown in Figure 9.5(*a*). Removal of oxygen to a composition below approximately $WO_{2.998}$ causes the structure to collapse along $\{102\}$ planes, thereby introducing $\{102\}$ *CS* planes into the crystal. The idealized structure of these defects is shown in Figure 9.5(*b*). They consist of blocks of four-edge-shared octahedra in a normal WO_3-like matrix. Slight reduction produces disordered $\{102\}$ *CS* planes, but as the composition approaches $WO_{2.95}$, these tend to become better ordered. When the *CS* planes are perfectly ordered, the composition of any phase is given by W_nO_{3n-1}, where n represents the number of octahedra separating the *CS* planes, measured along

(b)

(a)

(c)

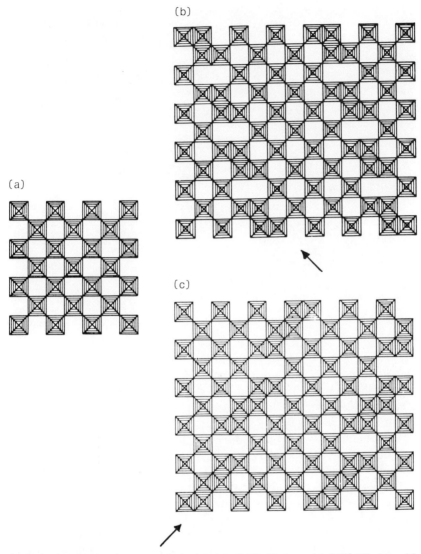

Figure 9.5 The idealized structures of (a), WO_3; (b), $\{102\}$ CS planes; (c), $\{103\}$ CS planes. The shaded squares represent WO_6 octahedra which are linked by corner sharing in WO_3 and by edge sharing within the CS planes. The arrows in (b) and (c) show the direction in which octahedra are counted in order to measure the value of n in the formulae W_nO_{3n-1} and W_nO_{3n-2} for the $\{102\}$ and $\{103\}$ series respectively.

the direction shown by an arrow in Figure 9.5(b). Such families of oxides are known as *homologous series*. The lower limit of the $\{102\}$ CS series is temperature-sensitive but corresponds to a composition of about $WO_{2.94}$, that is, an oxide of formula $W_{18}O_{53}$.

Further reduction, to take the composition below about $WO_{2.93}$, produces

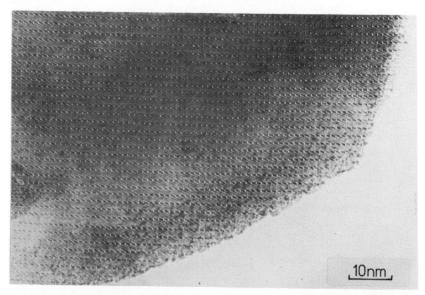

Figure 9.6 Electron micrograph of a $\{103\}$ CS phase. The dark blobs of contrast represent the groups of six edge-sharing octahedra which form the CS planes. Some defects in the CS plane ordering are also apparent. The spacing of the CS planes reveal that the material has composition close to $W_{18}O_{52}$.

a different CS structure in which blocks of six edge-shared octahedra form, as shown in Figure 9.5(c). The CS planes now lie on $\{103\}$ planes. These $\{103\}$ CS planes, which are usually fairly well ordered, persist alone over a composition range of approximately $WO_{2.92}$ to $WO_{2.87}$. Ordered $\{103\}$ CS planes will give rise to a homologous series of oxides with a general formula of W_nO_{3n-2}; the experimentally observed composition interval corresponding to CS phases in the range $W_{25}O_{73}$ to $W_{16}O_{46}$. Figure 9.6 shows an electron micrograph of the $\{103\}$ CS phase $W_{18}O_{52}$. In the composition region between $WO_{2.94}$ and $WO_{2.92}$, two-phase equilibrium occurs between $\{102\}$ and $\{103\}$ CS planes.

The CS behaviour of WO_3 when reduced is similar to that encountered when TiO_2 and MoO_3 are reduced. In general one finds that the CS plane type changes as the degree of reduction increases, and that ordered phases are more readily detected at greater degrees of reduction. In the TiO_{2-x} system, for example, lower degrees of reduction are accommodated on $\{132\}$ CS planes and greater degrees of reduction on $\{121\}$ planes, the changeover occurring at compositions between approximately $TiO_{1.93}$ and $TiO_{1.90}$.

In Figure 9.7 we show the structures of two phases from this system, $Ti_{16}O_{31}$ from the $\{132\}$ homologous series, and Ti_9O_{17} from the $\{121\}$ series. In these figures the anion packing has been emphasized rather than the metal coordination polyhedra as in Figure 9.5. This is to draw out the distinction between the organization of a CS phase and the occurrence of point

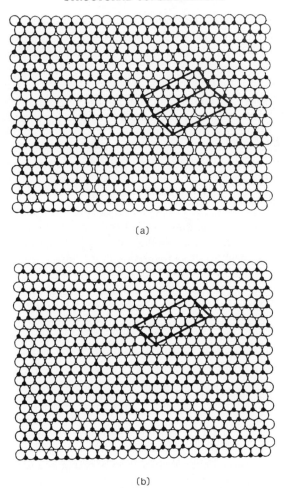

(a)

(b)

Figure 9.7 The structures of (a) $Ti_{16}O_{31}$, containing $\{132\}$ CS planes; and (b) Ti_9O_{17} containing $\{121\}$ CS planes, drawn so as to emphasize the packing of anions, shown as open circles, and cations, shown as filled circles. The full lines outline the unit cells of the structures, which, in (a), can be regarded as composed of two sub-cells.

defects. In Chapter 6 it was noted that the reduction of TiO_2 could be thought of as producing interstitial Ti point defects. Measurement of density would tend to confirm that interpretation of the data. However, examination of the structures in Figure 9.7 shows that, although extra Ti atoms have been added to the crystal, and that they occupy positions in the structure not normally occupied by metal atoms, a description in terms of interstitials is misleading. The ordering of the 'interstitials' is such that they are now integrated into the structure, with the consequence that the 'point defects' have been eliminated. If defects exist in these materials, they are disordered CS planes, that is *planar* rather than *point* defects.

Close examination of all the systems which rely upon CS to accommodate changes in composition reveals subtle degrees of complexity, particularly when reaction with other oxides is used to reduce the anion to cation ratio. If, for example, WO_3 is reacted with Nb_2O_5, CS planes form on $\{104\}$ and $\{001\}$ planes. These do not form in the binary tungsten–oxygen system and reveal that complex factors are involved in determining the planes upon which CS actually occurs. Similarly, if TiO_2 is reacted with Cr_2O_3 to produce a phase with a composition between approximately $(Ti, Cr)O_{1.93}$ and $(Ti, Cr)O_{1.90}$ a *swinging CS* region occurs. Within this composition range, any composition at all seems to have a unique structure. The CS plane spacings can vary, to accommodate different anion to cation ratios, of course, but in addition the indices of the CS plane can lie anywhere between $\{132\}$ at the oxygen-rich end of the phase range, to $\{121\}$ at the oxygen-poor end. It should be stressed that these phases are ordered, and that every composition prepared has an ordered arrangement of CS planes, with a definite spacing and CS orientation. If TiO_2 is reacted with the structurally similar oxide Fe_2O_3 instead, quite different CS structures form which swing from $\{020\}$ planes to $\{121\}$ planes.

It is clear, therefore, that each system must be treated on its merits, and composition is not the only factor which controls the structure of CS compounds, as temperature and impurities also have important roles to play. Moreover, it has been found that CS phases formed at high temperature often persist at lower temperatures, even in circumstances where they are thermodynamically unstable. The microstructures present in such phases therefore depends upon the history of the sample; a fact of some importance for both chemical reactivity and physical properties.

9.5.2 Chemical or unit cell twinning

Apart from CS planes a number of other types of planar boundary can alter the composition of a crystal; one such being the twin plane. The presence of a twin plane in a crystal is rather analogous to the presence of a mirror, and the two parts of the crystal united along the twin plane are mirror images of one another. In practice, not all twin planes produce composition changes, but many do. In this latter case, if a set of ordered twin planes exists in a crystal, it will produce, from the point of view of diffraction studies, a new phase. Homologous series of phases will result if the spacing between the twin planes can take a number of discrete values. If the twin planes are disordered we will have a non-stoichiometric phase with a measurable composition range. The phenomenon giving rise to changes in composition is referred to as *chemical twinning* (CT), or *unit cell twinning*.

We can illustrate this structural mode of incorporating changes in the metal to non-metal ratio in a crystal by referring to the $PbS–Bi_2S_3$ system. At the PbS-rich end of the composition range two phases have been characterized, *heyrovskyite*, $Pb_{24}Bi_8S_{36}$, and *lillianite*, $Pb_{12}Bi_8S_{24}$. Their structures, together

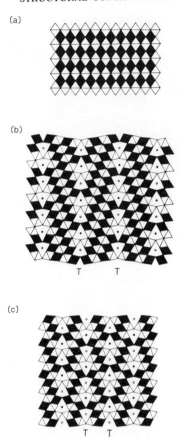

Figure 9.8 The structures of (a) PbS, projected on to (110); (b) lillianite, projected on to (001); and (c), heyrovskyite, projected on to (001). In each diagram the structures are shown as a packing of metal–sulphur octahedra; those at a higher level are shown in light relief and those at a lower level dark. The structures (b) and (c) can be regarded as made up of slabs of PbS structure joined along twin planes, marked T, which contain metal atoms in trigonal prismatic coordination.

with that of *galena*, PbS, are shown in Figure 9.8. It is seen that the structures of the minerals are composed of slabs of PbS, which has the *NaCl* type structure, united along twin planes, which lie on (311) planes with respect to the PbS unit cell. An electron micrograph of heyrovskyite, which reveals the twin repeat sequence clearly, is shown in Figure 9.9. In addition there are sites of a new geometry formed in the twin planes which do not exist in the original PbS structure. These may be filled with, for example, larger atoms, or may provide a pathway for enhanced chemical diffusion or reactivity. Should the width of the PbS slabs vary, we will generate other members of the heyrovskyite and lillianite series. Many such phases have, in fact, been found, particularly in mineral samples. In these, the ordering of the twin planes can be quite

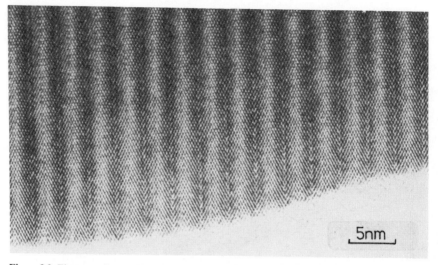

Figure 9.9 Electron micrograph of the chemically twinned phase heyrovskyite, built up of a series of galena-like slabs seven octahedra in width. Careful measurement reveals that some slabs are wider than normal, and are eight octahedra in width.

complicated, with some materials seeming to be composed of two alternating slabs of galena which are of different widths.

Chemical twinning occurs in a number of systems besides the lead–bismuth sulphides, and quite a wide range of compounds favours this mode of altering the anion to cation ratio in the parent structure. One rather elegant series, which forms in the Mg–Mn–B–O system, makes use of glide-twinning to accomplish the necessary changes in composition. These, and many other examples, will be found by consulting the sources listed in the Supplementary Reading section at the end of this chapter.

9.5.3 *Planar intergrowths*

In the CS and CT phases, the structures on each side of the planar boundary have been the same. It is quite easy to conceive of the situation where structures on each side of the boundary are somewhat different, in which case we meet with the phenomenon of *intergrowth*. Once again, this structural device was well known to mineralogists before its significance was appreciated in terms of non-stoichiometry.

There are a large number of materials which encompass a range of composition variation by way of intergrowth. Examples which have been described in the past include the hexagonal ferrites, the intergrowth tungsten bronzes and many silicate systems, particularly those related to the mineral pyroxene. Here we will mention three examples of intergrowth systems which have been characterized in recent years.

In the barium siliconiobates, intergrowth between $A_3M_6Si_4O_{26}$, where A can be Ba, Sr or 2K and M can be Nb or Ta, and $A_3Nb_{8-x}M_xO_{21}$, where A can be K or Ba and M can be Nb, Ti, Cr, Fe, Ni, Mg or Zn, has been found to occur. These structures consist of frameworks of MO_6 octahedra with the large K, Ba or Sr atoms occupying tunnels within the framework. The complexity of the system is enormous, and in the composition range between $Ba_3Nb_4Ti_4O_{21}$ and $Ba_3Nb_6Si_4O_{26}$, i.e. in the composition range corresponding to oxygen to metal ratios lying between 1.9091 and 2.0000, about 100 structures have been described. More would undoubtedly be discovered if the system was reinvestigated.

Another extensive intergrowth series is found in the $(Na, Ca)_nNb_nO_{3n+2}$ series of compounds. The structures of these phases, which are closely related to that of the mineral *perovskite*, are shown in Figure 9.10. In these materials slabs of perovskite structure are united along lamellae of rocksalt structure. The oxides are thus intergrowths of perovskite with CaO. Between compositions of $(NaCa_8)Nb_9O_{31}$, i.e. $MO_{4.5000}$ and $(Na_{12}Ca_{188})Nb_{200}O_{694}$, i.e. $MO_{4.2550}$, about 50 structures have been characterized, each of which is composed of various stackings of the $n = 4$ and $n = 5$ sequences shown in Figure 9.10. Clearly, any compositions made up between these two end members can yield an ordered stacking sequence to accommodate the stoichiometry exactly. Similar complexity exists between the other members of the series shown in Figure 9.10, so that over the whole of the phase region an enormous number of structures can be prepared.

This behaviour seems common in the perovskites. Besides the niobates just

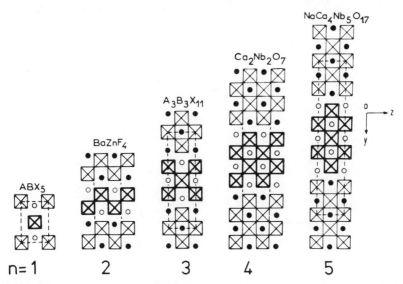

Figure 9.10 Idealized structures of the $A_nB_nO_{3n+2}$ series of perovskite related phases for values of n ranging from 1 to 5.

referred to, the bismuth titanates, bismuth tungstates, strontium titanates and many other related systems are known to show similarly complex behaviour. Many more examples are likely to be discovered in the perovskites as more systems are investigated in depth.

The third example to be mentioned here is that of the intergrowth tungsten bronzes. The structures of the parent phases, WO_3 and the hexagonal tungsten bronze structure, have already been illustrated in Figures 9.3 and 9.5. As has been pointed out, the hexagonal tungsten bronze structure is formed when potassium is reacted with WO_3, within the composition range of from $K_{0.19}WO_3$ to $K_{0.33}WO_3$. What happens when the amount of potassium is less than the minimum needed to form the hexagonal tungsten bronze structure?

(a)

(b)

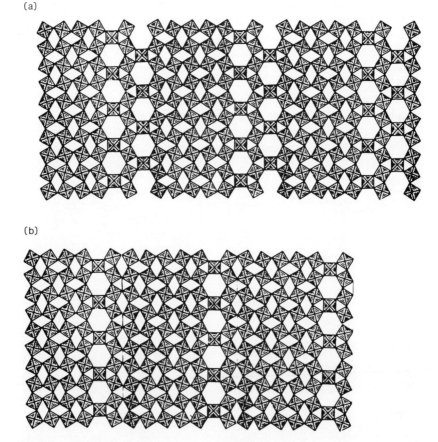

Figure 9.11 The idealized structures of two intergrowth tungsten bronze phases, (a) containing double rows of hexagonal tunnels, and (b) containing single rows of tunnels. The tungsten trioxide matrix is shown as shaded squares and the hexagonal tunnels are shown empty, although in the known intergrowth tungsten bronzes the tunnels contain variable amounts of metal atoms.

In this case an intergrowth between the hexagonal tungsten bronze structure and WO_3 is found. Usually the strips of hexagonal structure are two tunnels in width in the K_xWO_3 system, and this structure is illustrated in Figure 9.11(a), but single tunnels and wider strips have also been observed. Similar *intergrowth tungsten bronzes* are also known in a number of other systems, including Rb–, Cs–, Ba–, Sn–, and Pb–WO_3. In the non-alkali metal intergrowth bronzes, single tunnels seem to be preferred. This structure is shown in Figure 9.11(b), and a high resolution electron micrograph of a Ba intergrowth bronze which possesses this structure is shown in Figure 9.12.

In all these phases the separation of the strips of hexagonal tunnels increases as the concentration of the large interpolated atoms decreases, and homologous series of compounds form. Because the hexagonal tungsten bronze parent structure is able to tolerate a considerable range of composition due to variable filling of the hexagonal tunnels, it is likely that these intergrowth bronzes will also behave in the same way. In these phases, therefore, we have two ways of accommodating the change of composition, either by changing the relative numbers of hexagonal tunnels with respect to the WO_3 matrix, or else by varying the degree of filling of the hexagonal tunnels themselves.

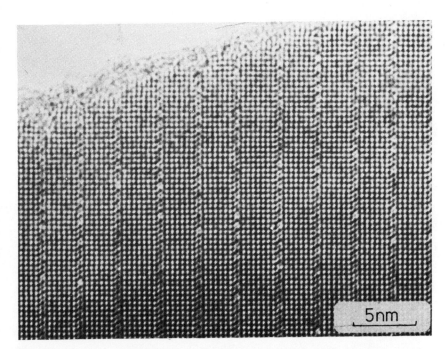

Figure 9.12 Electron micrograph of the intergrowth tungsten bronze phase Ba_xWO_3, showing the single rows of tunnels clearly. Each black spot on the image represents a tungsten atom, and many of the hexagonal tunnels seem to be empty or only partly filled with barium.

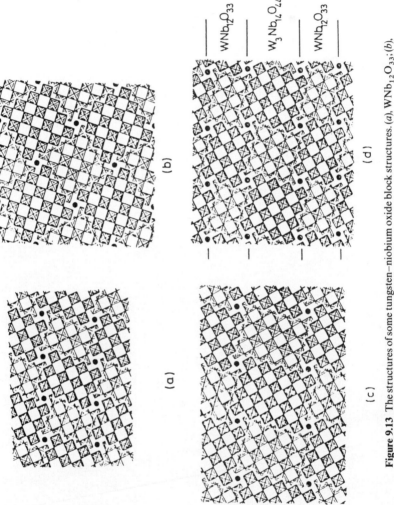

Figure 9.13 The structures of some tungsten–niobium oxide block structures. (*a*), $WNb_{12}O_{33}$; (*b*), $W_3Nb_{14}O_{44}$; (*c*), $W_5Nb_{16}O_{55}$; and (*d*), $W_4Nb_{26}O_{77}$, which is an ordered intergrowth of (*a*) and (*b*). The shaded squares represent MO_6 octahedra which are corner-sharing at the column centres and edge-sharing at the periphery of each column.

9.6 Three-dimensional faults and intergrowths

In the broadest sense, non-stoichiometric compounds containing point defect clusters, *CS* planes, *CT* planes or intergrowths all contain three-dimensional defects. There are, however, some crystallographically more surprising types of three-dimensional faults known to occur in oxides and sulphides, and for completeness, just two of these will be mentioned here.

The first is found in a group of niobium oxides close in composition to Nb_2O_5. In the binary system these oxides lie between the composition limits of $NbO_{2.5}$ and $NbO_{2.42}$, and in them two sets of *CS* planes intersect to divide the materials into columns of WO_3-like structure. The columns, in projection, look like rectangular blocks, and hence a common name for these materials is *block structures*. This system is one of many which was originally considered to consist of one non-stoichiometric phase but which has now been shown to consist of a multitude of phases. Indeed, the number is increased greatly by reacting Nb_2O_5 with other oxides, and we can pass from phases with composition below $NbO_{2.5}$, formed by reaction with TiO_2, for example, to compositions above $NbO_{2.5}$, by reaction with oxides such as WO_3. Some examples of the compounds which have been found to occur in the Nb_2O_5–WO_3 system are shown in Figure 9.13.

These changes of stoichiometry are brought about by changing the block sizes in the compounds. The dimensions of the rectangular cross-sections of the columns control the overall stoichiometry of the phase; the larger the dimensions, the nearer the composition comes to MO_3, and the smaller the blocks, the more the composition falls towards $MO_{2.0}$. Changes in local composition seem to be taken care of by local rearrangements in the block sizes and packings. An example of an incompletely reacted crystal which shows a variety of block sizes is shown in Figure 9.14.

Although it might be thought that reactions between these phases, or the interconversion of one block size to another would be difficult, quite the reverse is true. In fact the reactions of the block structures take place very rapidly. For example, reactions between Nb_2O_5 and WO_3 produce perfectly ordered phases within 15 minutes at about 1400 K, and reduced niobium oxides such as $Nb_{12}O_{29}$ oxidize at temperatures as low as 700 K. After such low-temperature oxidation reactions it is found that the block structure of the starting oxide is retained. This, however, necessarily has the wrong block sizes for the new composition. Annealing of the oxidized material at higher temperatures then produces a complex series of block rearrangements to adjust the block sizes to that proper for the more oxidized composition. This latter reaction, of course, takes place at constant composition. The reactions are coherent and the re-arrangement of the blocks seems to depend upon a desire to minimize geometrical misfit and hence probably elastic strain energy in the structure, and has led to the proposal that the reaction mechanisms involve minimum block reconstruction. These ideas clearly have something in common with ideas on topotaxy in solid-state reactions.

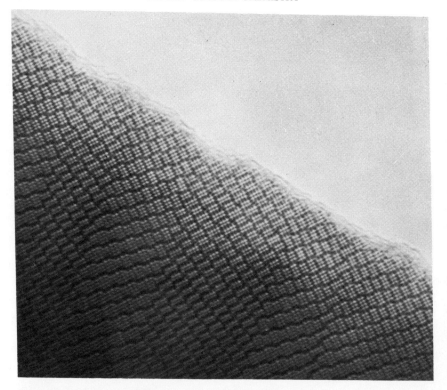

Figure 9.14 Electron micrograph of a disordered block structure phase formed in the Nb_2O_5–WO_3 system. The image clearly shows the block outlines. A number of different block sizes can be seen, each of which will correspond to a unique composition.

A second group of compounds which contain three-dimensional faults comprises the so-called *PC* structures. This is a fairly large group of metal oxides which have, as their basic structural motif, the *pentagonal column*. This consists of a pentagonal ring of five MO_6 octahedra, the tunnel so formed being filled with an alternating chain of oxygen and metal atoms to form a pentagonal column, as shown in Figure 9.15(*a*). It seems, at first sight, unlikely that such a unit could fit readily into a host structure, but nevertheless it is found that groups of *PC*s can coherently exist within a WO_3 type of matrix. As a number of oxides can adopt this structure or a distorted version of it, a wide range of *PC* phases can form. As in previous discussions, two alternatives exist. If the *PC* elements are perfectly ordered then we generate one or more structurally related homologous series of ordered phases. Some examples of these phases are given by the oxide Mo_5O_{14}, illustrated in Figure 9.15(*b*), the tetragonal tungsten bronze structure illustrated in Figure 9.3(*a*), and the oxide $Nb_{16}W_{18}O_{64}$ which is an ordered variant of the tetragonal tungsten bronze structure in which only a percentage of the available tunnel sites are filled, shown in Figure 9.15(*c*). If the *PC* elements in the host structure are disordered

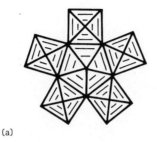

(a)

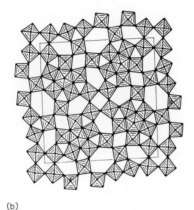

(b)

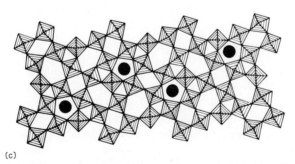

(c)

Figure 9.15 Some structures containing pentagonal columns: (*a*) an isolated pentagonal column; (*b*) the Mo_5O_{14} structure; and (*c*) the ordered tetragonal tungsten bronze structure of $Nb_{16}O_{18}O_{64}$, in which the square tunnels and some of the pentagonal tunnels are empty. The filled tunnels are indicated by filled circles.

a non-stoichiometric compound is generated. Such disorder occurs, for example, when WO_3 is reacted with Nb_2O_5 for short periods of time at temperatures below about 1500 K. An example of such a disordered crystal is shown in Figure 9.16.

In all of the *PC* phases, another form of non-stoichiometric variation is possible. In this case the O–M–O chains which occupy the pentagonal tunnels

H

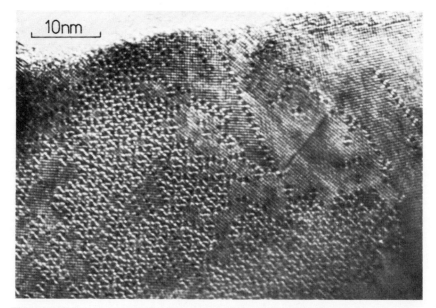

Figure 9.16 Electron micrographs of a disordered PC phase showing pentagonal columns, imaged as pairs of black blobs separated by white contrast, distributed largely at random in a WO_3 matrix.

can be incomplete, or else extra atoms can occupy tunnels which are normally empty. Interpolation is therefore also possible in these phases, thus matching the behaviour of the intergrowth tungsten bronzes discussed above.

9.7 Vernier structures

Vernier structures make use of a novel way to accommodate variations in anion to cation ratio which does not rely upon point defects at all. A number of these structures form in fluorite-related anion excess compounds, and we will illustrate this family by a consideration of the orthorhombic phases formed in the $YOF–YF_3$ system in the composition range between approximately $MX_{2.130}$ and $MX_{2.220}$.

The complexity of this system was unravelled by careful x-ray diffraction work. On powder photographs the strong reflections correspond to that of the fluorite type of cell, which is found in YOF. However, numerous other faint superstructure lines can also be seen on these films, indicating that a number of phases exist in the system. The positions of the lines denoting the new phases change almost imperceptibly on moving from one phase to another. Moreover, the lattice parameter of the cubic ZrOF subcell appears to change smoothly as the composition varies, and the system, at first sight, seems to fit a classical point defect description of a non-stoichiometric phase region. However, a careful interpretation of the x-ray results, in which no faint lines

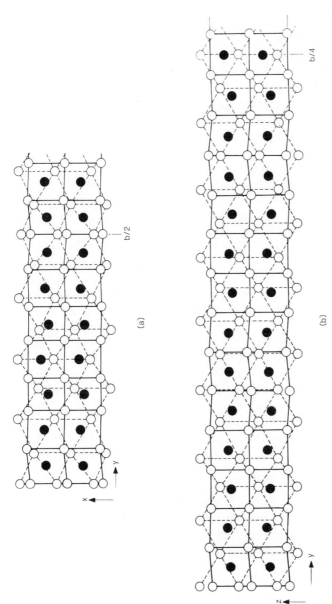

Figure 9.17 The structures of the vernier phases (a) $Y_7O_6F_9$ and (b) $Zr_{108}N_{98}F_{138}$ projected along [100]. The anion nets at $x/a = 0$ are shown as broken lines and those at $x/a = 1/2$ as full lines. The cations are shown as filled circles. If the projection of the anion nets is considered it is seen that a vernier relationship exists between the upper and lower layers.

are ignored and no unusual intensity changes are glossed over, yields a true picture of the real structural complexity present. Within the composition range of the non-stoichiometric region, every composition prepared has a different structure, and a large number of different ordered phases exist, forming a homologous series of compounds with a general formula $Y_nO_{n-1}F_{n+2}$.

The structures of these phases can be thought of as made up of unit cells of the fluorite type, of YOF in fact, stacked up in a sequence which is sometimes interrupted by some sort of ordered array of defects which allow extra anions to be incorporated into the structure. The nature of the defects responsible for this is remarkable. They are not anion interstitials or clusters, as in the case of UO_2, but defects in the regular arrays of anions which make up the structure. It is found that some of the anion nets which run approximately parallel to the (100) planes of the parent fluorite structure are changed from square form into hexagonal form. Now these latter nets contain a higher density of anions than the square nets, and this allows the extra atoms to fit into the structure readily. The change is shown in Figure 9.17 for the phase $Y_7O_6F_9$, which is the $n = 7$ member of the homologous series of phases $Y_nO_{n-1}F_{n+2}$, and the related compound $Zr_{108}N_{98}F_{138}$. The name *vernier structure* can now be appreciated. The positions of the metal atoms remain that of the fluorite parent, but the two sorts of anion arrays are in a vernier relationship to one another. The pitch of the vernier will vary from one compound to another in order to accommodate the correct anion to cation proportions in the structure.

Vernier structures are found in a number of other fluorite related systems, which include Zr–Nb–O and Zr–U–O. In these, the nets of anions, which are either square or hexagonal as we have described, do not always remain in the same sheet, but swap over regularly every half-unit cell. It is believed that this is to reduce strain energy in the structures. In addition, vernier structures are to be found in a number of other chemical systems, including some borides, carbides and silicides. Further information on these most interesting phases is given in the Supplementary Reading section at the end of this chapter.

9.8 Infinitely adaptive compounds

In the last few sections of this chapter we have come across instances of considerable structural complexity occurring within fairly narrow stoichiometry ranges. For the systems involved, any composition can be accommodated by an ordered structure. Examples that we have cited are the $YOF–YF_3$ phases, the $TiO_2–Cr_2O_3$ *swinging CS* phases, and the barium siliconiobates.

Here we mention another remarkable group of structures which form in the $Ta_2O_5–WO_3$ system. Recall first the crystal chemistry of Nb_2O_5 when reacted with WO_3. In this case a wealth of structures are created by fitting together rectangular columns. In the $WO_3–Ta_2O_5$ system at compositions near to

Ta_2O_5 it would be reasonable to believe that a similar situation would hold. In fact this is partly true, for a large number of compounds do form, but in this instance they are built up by fitting together pentagonal columns. The resultant phases are all structurally related to the low-temperature form of Ta_2O_5, and as in some of the other systems mentioned, every composition in the Ta_2O_5–WO_3 system would appear to produce a new ordered structure.

The structures of three of this large group of compounds are shown in Figure 9.18. It is seen that the idealized forms of these structures have a wavelike skeleton of PCs. As the composition varies, so the wavelength of this backbone changes, giving rise to the immense numbers of ordered structures found so far. In the real structures an extra degree of complexity arises due to the insertion of metal–oxygen octahedra into the PC nets, which results in a disruption of the regular waveforms. Nevertheless this is a very remarkable state of affairs, and J.S. Anderson coined the singularly apt name *infinitely adaptive compounds* for these and other phases which fall into this class.

Examples other than those listed in this chapter will be found in Anderson's

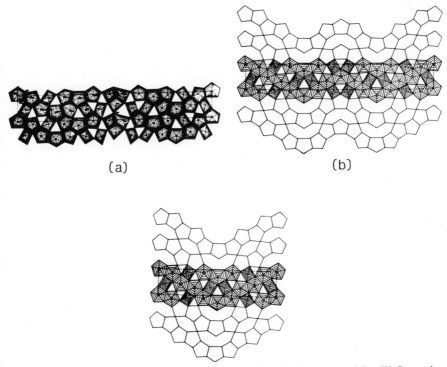

(a) (b)

Figure 9.18 The real structure of $Ta_{22}W_4O_{67}$ and the idealized structure of $Ta_{22}W_4O_{67}$ and $Ta_{30}W_2O_{81}$. These phases are built from pentagonal columns, shown as shaded pentagons, and octahedra, shown as shaded squares. The wavelength of the chains of pentagonal columns varies with composition in such a way that any given anion to cation ratio can be accommodated by an ordered structure.

paper (see Supplementary Reading). It is certain that the number of systems in which this type of behaviour is found will increase as more non-stoichiometric phase regions are studied in depth.

9.9 Some conclusions

This chapter concerns the relationship between stoichiometry and structure. It is not possible to do this topic any real justice in such a short space, and all that we have aimed to show is that naïve concepts of isolated point defects do not account for the experimentally-determined complexity shown in the structures of non-stoichiometric compounds. In the following chapter we will try to summarize the progress which is currently being made in an attempt to bring together the point defect concepts with which we began this book and the structural results just outlined.

9.10 Supplementary reading

Fortunately the subject matter of this chapter is well covered in a number of review articles. The classical article, by A.D. Wadsley, should be consulted before the others, but all contain different material and present different viewpoints of the subject.

A.D. Wadsley, in *Nonstoichiometric Compounds*, ed. L. Mandelcorn, Academic Press, New York (1963).

J.S. Anderson, in NBS Spec. Pub. 364, *Solid State Chemistry*, eds. R.S. Roth and S.J. Schneider, National Bureau of Standards, Washington (1972), 295.

J.S. Anderson, in *Defects and Transport in Oxides*, ed. M.S. Seltzer and R.I. Jaffee, Plenum, New York (1974).

J.S. Anderson, *J. Chem. Soc. Dalton Trans.* 1107 (1973).

B.G. Hyde, A.N. Bagshaw, S. Andersson and M. O'Keeffe, *Ann. Rev. Mater. Sci.* 4 (1974) 43.

B.G. Hyde, S. Andersson, M. Bakker, C.M. Plug and M. O'Keeffe, *Prog. Solid State Chem.* 12 (1979) 273.

E. Makovicky and B.G. Hyde, *Structure and Bonding* 46 (1981) 101.

R.J.D. Tilley, in *The Chemical Physics of Solids and their Surfaces*, 8, eds. M.W. Roberts and J.M. Thomas, Royal Society of Chemistry, London (1981).

The application of electron microscopy to non-stoichiometric compounds, and a comparison of electron microscopy to other diffraction techniques is covered in the review articles by L. Eyring and A.K. Cheetham in *Nonstoichiometric Oxides,* ed. O.T. Sørensen, Academic Press, New York (1981).

10 Thermodynamics and structure

10.1 Introduction

In the previous chapter we met a bewildering number of structures, and certainly have not catalogued all of the different ways in which non-stoichiometric variation can be accommodated in crystals. Our earlier ideas about non-stoichiometry therefore have to be revised in the light of these structural findings. It is the purpose of this final chapter to indicate some of the ways in which this may be accomplished, although the success which so far has been achieved is limited.

To some extent it is not difficult to outline the directions that can be taken. The central assumption in our discussion has been that the defects have been considered to be isolated. This means that interactions between defects have not been supposed to be of overriding importance. It has already been indicated in the text that defect interactions are important, and we made a few simple calculations to support this contention. The first extension to the theory that comes to mind is therefore to consider these interactions in more detail and perhaps to refine the calculations made earlier.

Such studies have been carried out and considerable progress has been made in this area. The work relies upon estimating the form of the potential between the atoms in the crystal and then minimizing the energy of the crystal and the defects contained therein. The procedure is complex and needs the computational power of modern computers for success, but nonetheless it is now a well-established method for the investigation of non-stoichiometric phases. Perhaps the most interesting findings of these studies have been those concerned with the structures of clusters of 'point' defects. These results will be referred to later.

A second area of study relates to thermodynamics and statistical mechanics. At various times in earlier chapters we have discussed both these disciplines. One way to improve on previous discussions is to consider in much greater detail the statistical thermodynamics of defect distributions. This has also been taken up with some success, and has led to a theory of non-stoichiometry in which microdomains of ordered structure exist in a macroscopically non-stoichiometric crystal.

Before considering either of these options further it is worth while to examine the idea of non-stoichiometry again.

10.2 Non-stoichiometry

As we have stated on several occasions, a non-stoichiometric phase is to be taken as a crystalline material which has a range of compositions open to it. We have excluded from this group solid solutions, although such a decision is to some extent arbitrary. This is because solid solutions do not usually possess properties or defect structures that are very different from that of the parent phases, whereas in the non-stoichiometric materials on which we have focused our attention both the properties and the defect structures have been significantly different from those of the parent phases. This seems a useful dividing line to take.

There is no reason why we should not retain this limitation and continue to concentrate upon phases which possess an authentic phase range due to varying populations of some sort of defects within the host structure. This means that we must exclude any compound that does not show a demonstrable composition range irrespective of its formula. That is, a material that has a complex formula obviously not in accord with Dalton's laws, such as $W_{18}O_{49}$, is not to be regarded as non-stoichiometric simply because it does possess a complex formula. Hence many of the line phases that we met with in the last chapter are to be considered as stoichiometric despite their unusual and frequently rather bizarre compositions.

When considering whether a compound is to be regarded as non-stoichiometric or not, it is important to consider the temperature at which the observations are made. Phases which are line phases at low temperatures may not retain this inflexibility at higher temperatures. Thus spinel, $MgAl_2O_4$, which we used in Chapter 4 as an example of a stoichiometric phase, loses this attribute as the temperature increases. A more complete phase diagram for the $MgO-Al_2O_3$ system, as shown in Figure 10.1, indicates that at high temperatures spinel can dissolve quite appreciable quantities of Al_2O_3, to show a considerable existence range. In general, therefore, we have to be on the look out for the effects of temperature and especially for a change from stoichiometric line phases at one temperature to a broad non-stoichiometric phase at a higher temperature.

It is also important to remember that the structural organization of the high-temperature phase may not be related to that of the low-temperature line phase. For example, suppose that the line phase is made up of an ordered array of planar boundaries in the parent structure. The high-temperature structure could simply consist of the same boundaries distributed in an ordered or disordered fashion within the bulk of the crystal. If the number of planar boundaries remains the same as at lower temperatures, the composition will also remain unchanged. On the other hand it is also possible that the planar

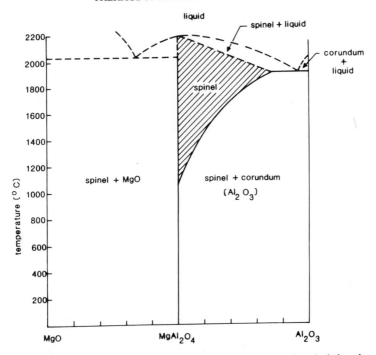

Figure 10.1 The MgO–Al$_2$O$_3$ phase diagram, showing the way in which the spinel phase becomes non-stoichiometric at higher temperatures. It is seen that the composition varies only on the Al$_2$O$_3$ side of the phase and not on the MgO side. The spinel phase field is shown shaded.

boundaries could 'dissolve' at higher temperatures to yield a crystal containing, perhaps, a disordered population of defect clusters with a variable composition.

With respect to this latter point it is germane to ask whether thermodynamics can clarify the situation, recalling that the definition of a non-stoichiometric compound was that such phases show bivariant behaviour. In the main, experiments to measure the way in which composition varies with partial pressure of the components present in the phase under examination are made at high temperatures. In principle, therefore, it should be possible to differentiate between a stoichiometric line phase and a non-stoichiometric phase simply by checking this aspect of thermodynamic behaviour. This would be doubly useful, in fact, as the structural studies are invariably carried out at room temperature, so by comparing the room-temperature structural results with the high-temperature thermodynamic data we should be able to resolve the problem of a change of defect type with temperature.

What problems are encountered in practice when this is attempted? When the schematic diagrams shown in Figures 8.5 and 8.7 are considered, it appears to be simple to differentiate between a two-phase region and a single-phase region. However, this is not necessarily so. In Figure 10.2 we show the way in

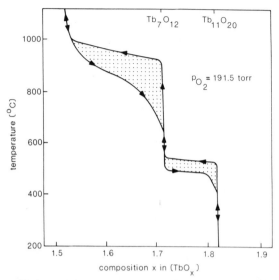

Figure 10.2 An oxidation–reduction curve for the oxide system Tb–O obtained at a constant oxygen pressure of 2.25×10^4 Pa (191.5 torr). The arrows indicate the paths followed during oxidation and reduction, which are reproducible, and not coincident. Data reproduced from Hyde and Eyring (1965) (see Supplementary Reading).

which the composition of a sample of $Tb_{11}O_{20}$, that is, $TbO_{1.8182}$, changes with temperature at a fixed oxygen pressure. An examination of the reduction path seems to indicate that we have two stoichiometric line phases, $Tb_{11}O_{20}$ and Tb_7O_{12}, and that the Tb_2O_3 phase is non-stoichiometric and oxygen-rich, with a composition of $TbO_{1.5+x}$. On re-oxidation, however, quite different behaviour is found which is not so easily interpreted. Clearly many regions of the curves are neither horizontal or vertical, which indicates bivariant behaviour quite at variance with the reduction cycle. The problem is compounded by the fact that these curves are quite reproducible, and so cannot be dismissed as indicating that a non-equilibrium situation holds.

Now suppose that we have some structurally complex phases present, as the formulae $Tb_{11}O_{20}$ and Tb_7O_{12} suggest, and that these phases contain differing numbers of ordered 'defects' in the parent TbO_2 phase. Clearly reduction, which involves putting in more 'defects', will require a different mechanism than oxidation, which will involve removal of the 'defects'. There is no *a priori* reason why these two processes should take place at the same rate and so, in cases involving a series of microphases, *hysteresis*, as shown on Figure 10.2, would be expected to be the rule.

We can usefully consider this in a little more detail. In Figure 10.3 we reproduce some very precise data showing the oxidation and reduction data for the rutile form of TiO_2. It is clear that there is considerable hysteresis and it is very difficult to be precise about the number of phases present in the

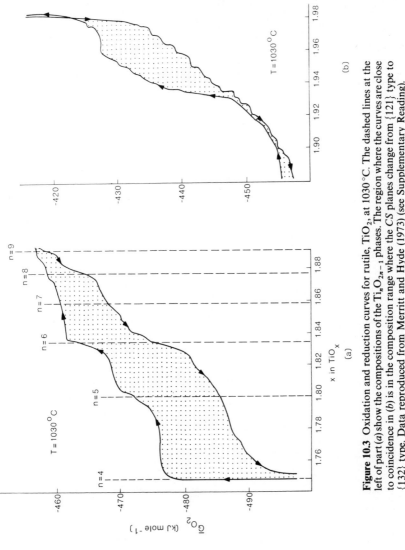

Figure 10.3 Oxidation and reduction curves for rutile, TiO_2, at 1030 °C. The dashed lines at the left of part (a) show the compositions of the Ti_nO_{2n-1} phases. The region where the curves are close to coincidence in (b) is in the composition range where the CS planes change from {121} type to {132} type. Data reproduced from Merritt and Hyde (1973) (see Supplementary Reading).

composition range spanned, let alone whether the behaviour at any one point should be classified as univariant or bivariant. Fortunately, the structures of the phases occurring in the system are well known. The composition range between TiO_2 and Ti_4O_9 is spanned by a series of crystallographic shear phases with a series formula of Ti_nO_{2n-1}. In the lower composition region the structures of Ti_4O_7, Ti_5O_9, Ti_6O_{11}, Ti_7O_{13}, Ti_8O_{15} and Ti_9O_{17} all contain ordered arrays of crystallographic shear planes on $\{121\}$ planes, as shown in Figure 9.7(b). In the composition range between $Ti_{16}O_{30}$ and TiO_2 the crystallographic shear planes lie upon $\{132\}$ planes, as shown in Figure 9.7(a). Between these two regions, at a composition of about $TiO_{1.91}$, the crystallographic shear planes swing from one orientation to the other to form an infinitely adaptive phase range.

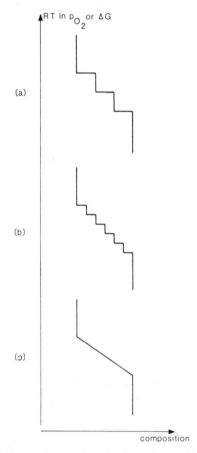

Figure 10.4 An idealized diagram showing that as the number of phases in a composition range increases, the expected free energy v. composition curves approach the continuous slope expected from a non-stoichiometric phase. The ability of the experiment to differentiate between the situations shown in (b) and (c) is of the greatest importance in practice.

The process of introducing and ordering planar boundaries will un-doubtedly be quite different to the process of removing and re-ordering the remaining planar boundaries. Hence it is hardly surprising that interpretation of the thermodynamic data is so difficult. Indeed, the thermodynamic data cannot be interpreted satisfactorily unless the structural information is also available.

We see, then, that the problem of interpretation of thermodynamic data is not dissimilar to the problem of interpretation of structural data. The precision of the interpretation will depend upon the precision of the technique. In a system containing a homologous series of compounds it may be impossible to differentiate, in practical terms, between a bivariant region and a closely spaced series of univariant equilibria. Figure 10.4 shows this schemati-cally. When the added problem of hysteresis is taken into account it becomes clear that thermodynamic measurements alone will only rarely be able to define a system containing non-stoichiometric phases completely.

10.3 Ordering, assimilation and elimination of defects

In this section we want to link the structural results of the last chapter with the point defect ideas regarding the causes of non-stoichiometry noted in Chapter 4. With this in mind, we will reiterate in summary form the 'point defect' modes of changing the anion to cation ratio in a crystal that we have already used. These are as follows.

(i) *Interpolation.* In interpolation, extra atoms are introduced into the structure in positions that are normally unoccupied in the parent phase. A good example of interpolation is provided by the non-stoichiometric phase $Ti_{1+x}S_2$. Other instances are given in the previous chapter. The point defects relevant to interpolation are *interstitials*.

(ii) *Subtraction.* Subtraction simply means that some of the atoms that should be present in the structure are missing. If only a few are absent then the departure from the stoichiometric ratio will be small, and this will be the normal case. An example of a non-stoichiometric phase employing subtrac-tion would be $Ni_{1-x}O$, and the defects involved would be *vacancies*.

(iii) *Substitution.* In this case, atoms of one type are substituted for those of another type in the structure. If the atom types involved in the substitution are of the same valence then we have simple solid solution formation, as in $Ni_xMg_{1-x}O$. The substitution can be more complex and yet still retain this solid solution feature. For example, we mentioned the sulphide phase lillianite, $Pb_{12}Bi_8S_{24}$, in the previous chapter. It is found that some of the Pb atoms can be replaced by a combination of Ag and Bi, as two Pb atoms are equivalent, in charge terms, to one Ag plus one Bi. This produces the phase *gustavite*,

$Pb_4Ag_4Bi_{12}S_{24}$. This type of substitution is very commonly encountered in mineral crystals, where it is referred to as *isomorphous substitution*. Despite this complexity, we are still dealing with normal solid solution formation. On the other hand, if the substituting ions are of the 'wrong' valence, or if the correct balance of numbers is not preserved, then a non-stoichiometric phase may be formed. Lithium doped nickel oxide, $Li_xNi_{1-x}O$, provides an example. In this case, the oxygen sub-structure remains intact but vacancies form in the metal atom array. Clearly there is no one sort of defect associated with substitution, as the nature of the substitution will control the nature of the compensating entities required.

The structural conclusions presented in the previous chapter suggest that ordered or disordered aggregates of defects are the rule rather than the exception in non-stoichiometric crystals. It is therefore rather useful to try to present a unified scheme to bring the structural chemistry and defect chemistry together. There are three important variables to consider: the interactions between the defects, the temperature and the structure of the non-stoichiometric phase. These are shown as three axes in Figure 10.5.

These variables are not, of course, independent of one another, and nor are they the only ones that could be selected. The interaction between the defects can be thought of in terms of free energies, or, more properly, in terms of the balance between enthalpy and entropy. Random arrangements of defects implies a high entropy and weak interactions, while ordered arrays of defects means that the entropy contribution is small, the enthalpy of the interactions is high and the phase is likely to be a stoichiometric compound.

One slice through the three-dimensional manifold is shown in Table 10.1.

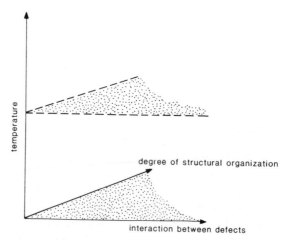

Figure 10.5 Schematic representation of the parameters which are important in determining the structure of a non-stoichiometric phase. The shaded sheet is an indication of the section of this three-dimensional manifold shown as Table 10.1.

Table 10.1 A summary of defect organization and structure in non-stoichiometric phases

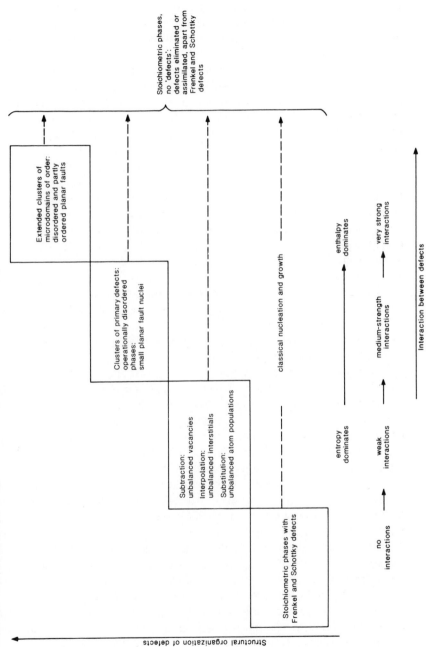

This must be regarded not as a fixed plane, but rather as a mapping of the possibilities on one sheet. At the left of the diagram the situation in normal stoichiometric materials is considered. In such compounds, Frenkel and Schottky defects are found. The interaction between these defects is supposed to be very weak indeed. Entropy effects are dominant and enthalpy effects are negligible in this case.

As we move to the right we come to non-stoichiometric phases with vacancies, interstitials, or substituted atoms. If the interactions between these defects are still considered to be weak then they will be distributed at random, and entropy will still dominate enthalpy. It is very uncertain whether such a situation will exist in any real compound, as defect interactions will certainly become dominant at very low defect concentrations. In this case, the structural situation will demand clusters of defects, and this level of association is shown in the next column of the table. The nature of these clusters will depend upon the system under consideration, and may contain, for example, small crystallographic shear plane nuclei. The phase will be non-stoichiometric in an operational sense, which means that experimentally the material will behave exactly as the somewhat idealized non-stoichiometric materials that we have already considered in the early parts of this book. However, we are specifying a degree of ordering at a microscopic level much beyond that considered implicitly earlier. As the interactions between the defects increases, so the level of organization between the defect clusters will increase. The structural picture will now be of *microdomains of order* within the crystal matrix. As an example we can picture partly-ordered groups of planar faults in the phase. At this juncture enthalpy will dominate entropy.

If one imagines the interactions to be so great that the defects become completely ordered then they will be *totally assimilated* into the structure and no defects as such will be present. Thus, fully ordered Ti_5O_9 will be a defect-free stoichiometric phase. It may contain either Frenkel or Schottky defects, of course, but these will not change the composition at all. We have returned to the same situation that we encountered at the left of the table and so could imagine the sheet to be wrapped around a cylinder.

Although such a scheme provides an attractive summary of possibilities, it is unlikely that the sequence from left to right will be followed by any one material. If the interactions between defects are weak, then only the left side of the chart will be of relevance. If the interactions are strong, then we will pass directly to microdomains or to a new ordered phase. This latter situation will correspond to classical nucleation and growth of a second stoichiometric phase, and so takes us out of the present topic and into that of phase transformations. In the same way, an increase of temperature will tend to decrease interactions and be equivalent to emphasizing the entropy factor as against the enthalpy. The scheme outlined in Table 10.1 is thus to be taken simply as a guide to possibilities, and not as a rigid framework.

10.4 Theories and calculations

Structural studies such as those summarized earlier have demonstrated convincingly that randomly-distributed point defects are not to be found in grossly non-stoichiometric phases. Because of interactions, the defects become ordered, and form clusters of a new structure within the disordered non-stoichiometric phase. If the degree of ordering is high, the defects will be assimilated within microdomains of order, which will be ordered regions in which defects (in the point defect sense) no longer exist. In this case the defects present will, in fact, be the surfaces of the microdomains themselves. Should the microdomains become large, turning into macrodomains, then we revert to a classical two-phase system.

One way of thinking about microdomains in a disordered non-stoichiometric compound is to regard them as partly-ordered collections of atoms existing over a certain percentage of the volume of the solid. The partly-ordered atoms are bound together by interactions. At higher temperatures at least, the interactions will be competing with disordering tendencies, so that the ordered microdomains must be thought of as forming and redissolving continuously to yield a dynamic picture of the phase. To take the matter further theoretically, we can therefore concentrate either upon the statistical mechanics of the microdomains or else upon the interactions which give rise to the defect clusters which make up the microdomains. Both of these alternatives have formed the basis for extensive research programmes in recent years. The studies are at a fairly advanced level and so will be mentioned only briefly here. References are, however, given in the Supplementary Reading section at the end of the chapter for those who wish to pursue the subject further.

Statistical mechanical theories tend to start with point defect concepts similar to those presented earlier in this book. however, in contrast to our initial suppositions, the energy of each defect is now supposed to be a function of its position within the crystal. In order to define the statistical thermodynamics of this complex system it is then necessary to specify the nature of the interactions between the defects and how they change with position. It is very difficult to define this interaction for other than nearest-neighbour atoms or defects, and the interaction itself is often considered to be simply attractive or repulsive. The entropy may be taken into account by assuming a random distribution of defects as we outlined in Chapter 1 of this book, or else, if defects are considered to repel each other, by assuming that some sort of *site exclusion rules* have to be obeyed. We have seen that there is a strong connection between the degree of ordering, the entropy, and the interaction between defects, the enthalpy term, and it is this interdependence which proves difficult to express in mathematical terms. Despite such problems several notable theories to account for microdomain formation have emerged in recent years, and these have had some success in accounting for the behaviour of materials which appear to contain point defect clusters. The topic can be taken further, by those interested, by reference to the Supplementary Reading section at the end of this chapter.

Calculations have been made for two principal reasons, to elucidate the formation energies of defects and to clarify the nature of the interactions between defects. To calculate defect formation energies an ionic model is usually chosen, and coulomb-type interactions of the sort we have used earlier are regarded as dominant. The polarization of ions can be included in the calculations, and these effects are often of considerable importance. The calculations are lengthy and involve much computation. Besides calculating formation energies as such, the calculations can be used to compare the energetics of various models of defect structures. This has proved to be particularly useful for non-stoichiometric phases which contain defect clusters. It is difficult to determine the structure of a defect cluster from conventional diffraction techniques, but diffraction results often provide the basis for the construction of one or two reasonable models. Energy calculations can then be used to determine which of the possible configurations has the lowest energy and is the most likely to be present in the crystal. The forms of the defect clusters in $\approx$ FeO illustrated in the previous chapter were derived in this way.

The second aim of calculations has been to try to account for the way in which the defects order. In the main, the studies have concentrated on planar boundaries such as crystallographic shear planes, which order over distances of several tens of nanometres. Elastic strain energy seems to account well for this ordering, and may well play a significant role in controlling the microstructures of non-stoichiometric phases which make use of planar defects to adjust the anion to cation ratio. Further calculations are needed, however, before the general validity of this statement is proven.

10.5 Conclusions

In this book we have considered defects in both stoichiometric and non-stoichiometric compounds, some theories, largely thermodynamic in nature, to account for these defects and the consequences that the defect population has for some of the properties of these materials. The level chosen has been introductory in nature, intended to provide a basis for further study. To this end the Supplementary Reading sections, especially those in the last two chapters, list more advanced sources for this purpose and will provide a starting point for future thought.

10.6 Supplementary reading

The following references cover recent or significant advances in the study of non-stoichiometric compounds, especially the relationships between thermodynamics and structure. Most work is concerned with oxide chemistry which reflects the current situation in this area of study.

Two books which contain a collection of advanced review articles are E. Rabenau (ed.), *Problems of Nonstoichiometry*, North-Holland, Amsterdam (1970); O.T. Sørensen (ed.), *Nonstoichiometric Oxides*, Academic Press, New York (1981).

The whole topic is reviewed succinctly by D.J.M. Bevan, Chapter 49 in *Comprehensive Inorganic Chemistry*, Vol. 4, ed. A.F. Trotman-Dickenson, Pergamon, Oxford (1973).

The relationships between structure and thermodynamics are set out clearly by J.S. Anderson, in NBS Spec. Pub. 364, *Solid State Chemistry*, eds. R.S. Roth and S.J. Schneider, National Bureau of Standards, Washington (1972); *The Chemistry of the Solid State*, ed. C.N.R. Rao, Marcel Dekker, New York (1974); and also in *Problems of Nonstoichiometry* (E. Rabenau, ed.) *op. cit.*, p. 1.

Statistical thermodynamic theories are discussed by L. Manes, and calculations of the structures of defect clusters by C.R.A. Catlow, in *Nonstoichiometric Oxides* (O.T. Sørensen, ed.) *op. cit.* pp. 61, 100, respectively.

An example of the calculation of elastic strain energy and its role in the ordering of crystallographic shear phases is given by E. Iguchi and R.J.D. Tilley, *Phil. Trans. Roy. Soc. London* **A268** (1977) 55.

The results on the Tb–O system shown in Figure 10.2 are taken from B.G. Hyde and L. Eyring, *Rare-earth Research*, Vol. 3, L. Eyring (ed.), Gordon and Breach, New York (1965).

The results on the Ti–O system shown in Figure 10.3 are from R.R. Merritt and B.G. Hyde, *Phil. Trans. Roy. Soc. London* **A274** (1973) 627–661.

Index